PRAISE FOR
The Boys of Pointe du Hoc

"This insightful book effectively interweaves the story of D-Day . . . with the story of Reagan's presidency . . . as well as Reagan's legacy. Brinkley chose his subject wisely and argues his case convincingly."
—*Washington Post Book World*

"The boys of Pointe du Hoc are needed today. Douglas Brinkley's fine historical exposition weaves the courage of the American Rangers at D-Day into the fabric of the Reagan presidency to illuminate just what qualities are needed now to keep the United States strong and free. An important and entertaining book."
—Bill O'Reilly, host of *The O'Reilly Factor*

"In *The Boys of Pointe du Hoc,* historian Douglas Brinkley proves again his instinct for our strongest history, his ear for the music—as well as lyric—of our proudest World War II moment. Brinkley knows there is sometimes a theater to war and always to its remembrance. More than a grateful taps for those who gave so much on the cliffs of Normandy, his book is a bugle call of reveille for what they did."
—Chris Matthews,
host of *Hardball* and
author of *Kennedy and Nixon*

"Brinkley's book is an endearing tribute. . . . [I]nteresting."
—*Richmond Times-Dispatch*

"In this jewel of a book, Douglas Brinkley proves his skills as a master storyteller. With solid research and superb writing, he weaves together two dramatic events. The memorable battle of Pointe du Hoc is re-created in vivid detail, as is the story of the moving commemorative speech Ronald Reagan delivered at the site that helped launch a renewed appreciation of World War II veterans. Brinkley's idea of linking these two events yields a fascinating and original book."
—Doris Kearns Goodwin,
author of *No Ordinary Time*

"Compelling reading." —*New Orleans Times-Picayune*

"Riveting." —*Chicago Tribune*

"Doug Brinkley does a glorious job weaving together the story of Ronald Reagan's visit to Normandy and the U.S. Army Rangers who fought on D-Day. The result is a powerful tale that celebrates, and explores, the patriotism and pride inspired by America's brave soldiers." —Walter Isaacson, author of *Benjamin Franklin: An American Life*

"Brinkley has a fine style and captures the essence of the battle very well. . . . Well written." —*Philadelphia Inquirer*

"A dramatic recounting of the heroic exploits of the Army's 2nd Ranger Battalion on D-Day, June 6, 1944."
—*Atlanta Journal-Constitution*

"An excellent read for those interested in media popularization and political demagoguery. . . . [*The Boys of Pointe du Hoc*] is a most useful and readable case study of the making of popular history." —*Booklist*

"No one would remember World War II . . . writes prolific historian Brinkley, if it had not been for two speeches Reagan gave in Normandy on June 6, 1984. . . . He makes a solid case."
—*Kirkus Reviews*

Chris Selzer

About the Author

DOUGLAS BRINKLEY is a Clark Professor of history and director of the Theodore Roosevelt Center for American Civilization at Tulane University. His publications include *The Great Deluge*; *Parish Priest: Father Michael McGivney and American Catholicism* with Julie M. Fenster; the *New York Times* bestseller *Tour of Duty: John Kerry and the Vietnam War*; *Wheels for the World: Henry Ford, His Company, and a Century of Progress*; *Driven Patriot: The Life and Times of James Forrestal* with Townsend Hoopes; and *The Mississippi and the Making of a Nation* with Stephen A. Ambrose. He lives in New Orleans, Louisiana, with his family.

ALSO BY DOUGLAS BRINKLEY

The Great Deluge:
Hurricane Katrina, New Orleans, and the Mississippi Gulf Coast

Parish Priest: Father Michael McGivney and American Catholicism
(with Julie M. Fenster)

Tour of Duty: John Kerry and the Vietnam War

Windblown World:
The Journals of Jack Kerouac, 1947–1954

Wheels for the World: Henry Ford, His Company,
and a Century of Progress, 1903–2003

The Mississippi and the Making of a Nation
(with Stephen E. Ambrose)

American Heritage History of the United States

The Western Paradox: The Bernard DeVoto Reader
(editor, with Patricia Limerick)

Rosa Parks

The Unfinished Presidency: Jimmy Carter's Journey Beyond
the White House

John F. Kennedy and Europe (editor)

Rise to Globalism: American Foreign Policy Since 1939, eighth
edition (with Stephen E. Ambrose)

The Majic Bus: An American Odyssey

Dean Acheson: The Cold War Years, 1953–1971

Driven Patriot: The Life and Times of James Forrestal
(with Townsend Hoopes)

FDR and the Creation of the U.N.

THE BOYS OF POINTE DU HOC

RONALD REAGAN, D-DAY, AND THE U.S. ARMY 2ND RANGER BATTALION

DOUGLAS BRINKLEY

HARPER PERENNIAL

NEW YORK • LONDON • TORONTO • SYDNEY

HARPER PERENNIAL

A hardcover edition of this book was published in 2005 by William Morrow, an imprint of HarperCollins Publishers.

FIRST HARPER PERENNIAL EDITION PUBLISHED 2006.

Designed by Renato Stanisic

The Library of Congress has catalogued the hardcover edition as follows:

Brinkley, Douglas.
The boys of Pointe du Hoc : Ronald Reagan, D-Day, and the U.S. Army 2nd Ranger Battalion / Douglas Brinkley.—1st ed.
p. cm.
ISBN 0-06-056527-6
1. World War, 1939–1945—Campaigns—France—Normandy—Anniversaries, etc. 2. United States. Army. Ranger Battalion, 2nd—History. 3. World War, 1939–1945—Regimental histories—United States. 4. Normandy (France)—History—Anniversaries, etc. I. Reagan, Ronald. II. Title.

D756.5.N6B748 2005
940.54'2142—dc22 2005040008

ISBN-10: 0-06-056530-6 (pbk.)
ISBN-13: 978-0-06-056530-5 (pbk.)

06 07 08 09 10 ❖/RRD 10 9 8 7 6 5 4 3 2 1

To D-Day historian Ronald J. Drez

and the

men of the U.S. Army 2nd Ranger Battalion during World War II

These are the boys of Pointe du Hoc. These are the men who took the cliffs. These are the champions who helped free a continent. These are the heroes who helped end a war.

—Ronald Reagan,
Speech at Pointe du Hoc, Normandy, June 6, 1984

Almighty God: Our sons, pride of our Nation, this day have set upon a mighty endeavor . . . to set free a suffering humanity.

Lead them straight and true; give strength to their arms, stoutness to their hearts, steadfastness in their faith.

They will need Thy blessings. . . . They will be sore[ly] tried, by night and by day, without rest—until the victory is won. The darkness will be rent by noise and flame. Men's souls will be shaken with the violence of war.

For these men are lately drawn from the ways of peace. They fight not for the lust of conquest. They fight to end conquest. They fight to liberate. They fight to let justice arise, and for tolerance and good will among all Thy people. They yearn but for the end of battle, for their return to the haven of home.

Some will never return. Embrace these, Father, and receive them, Thy heroic servants, into Thy kingdom.

And for us at home—fathers, mothers, children, wives, sisters, and brothers of brave men overseas—whose thoughts and prayers are ever with them—help us, Almighty God, to rededicate ourselves in renewed faith in Thee in this hour of great sacrifice. . . .

Thy will be done, Almighty God.
Amen.

—President Franklin D. Roosevelt, D-Day prayer to the nation, June 6, 1944

CONTENTS

THE BOYS OF POINTE DU HOC

Grave of a 2nd Battalion Ranger killed on D-Day. (U.S. Army Military History Institute)

Introduction

SETTING THE STAGE

One brisk morning at the White House in early October 1981, as President Ronald Reagan was preparing for a state visit from French president François Mitterrand, his deputy chief of staff, Michael Deaver, informed him that the French government was eager to present him a prestigious decoration when the two leaders met in historic Yorktown, Virginia. A quizzical Reagan was baffled. "What decoration?" An unsure Deaver answered, "I think it's the Croix de Guerre." Clearly displeased by the rejoinder, Reagan, his face flushing, balked. His lips, in fact, contorted into a half pout. What was this nonsense? He had not earned epaulets. There was no Purple Heart or Bronze Star in his attic trunk that could be dusted off and flaunted for the grandkids. He was no colossal World War II hero. His tame war years were spent stateside in San Francisco and Culver City, feigning to be a bomber pilot, directing and starring in over three hundred training films for the Army Air Corps. "But that's for bravery," Reagan impatiently snapped. "All I ever did was fly a desk. We'd better get this straightened out right away. I couldn't possibly accept the Croix de Guerre."

Evan Galbraith—realizing that Reagan was aghast at the implication that he was qualified for such a distinguished military award—quickly looked into the matter. Getting on the telephone, Galbraith, soon to become the Reagan administration's second ambassador to France, discovered there was indeed a bit of a mix-up. "It's not the Croix de Guerre they'll be giving you after all," a relieved Galbraith blurted out. "It's the Légion d'Honneur." A still puzzled Reagan asked for what possible reason. "Statesmanship," Galbraith answered.

Suddenly Reagan's whole demeanor changed. A relaxed, confident smile now beamed forth. The look of consternation had dissipated from his furrowed brow. His usual radiant, genial mood was back in full force. Comically, he pretended to be fixing his tie knot. "I can play *that* role," he laughed.

Indeed he could. This story took place on the eve of celebrating the two hundredth anniversary of France's coming to the aid of George Washington's Continental Army at the Battle of Yorktown. President Reagan reciprocated in June 1982, when he visited France as part of his first European trip since becoming president.* Along with his devoted wife

*During this European tour he also visited the Vatican, London, Bonn, and Berlin.

Nancy, the First Lady, he lodged at both the ornate Grand Trianon, Louis XIV's retreat at Versailles, still awe-inspiring after three centuries, and at the Paris residence of Ambassador Galbraith. Keeping a hectic schedule, Reagan held substantive meetings with President Mitterrand, Paris mayor Jacques Chirac, Japanese prime minister Zenko Suzuki, and British prime minister Margaret Thatcher at a three-day world economic summit in the Salle du Congrès at the Palace of Versailles. Whenever import-export issues were raised or Middle East politics thrashed out, Reagan invariably steered the conversation back to Cold War geopolitics. "We . . . are moving forward to restore America's defensive strength after a decade of neglect," Reagan exhorted at dinner on June 3. "A strong America and a vital, unified alliance are indispensable to keeping the peace now and in the future, just as they have been in the past."

With his "Peace Through Strength" foreign policy approach, Reagan was trying to create the indelible image of the tough, protean Cold War statesman, a trifecta combination of FDR (optimist/communicator), Truman (straightforward/honest), and Eisenhower (amiable/shrewd). Throughout his European travels he constantly evoked the need to reenergize the grand anti-Fascist alliance of 1941–1945, which led to victory in World War II. Only this time around, Reagan believed, the global democratic crusade had a new enemy: the Soviet Union (a.k.a. the Evil Empire). On June 5, Reagan gave a well-rehearsed radio address from the Palace of Versailles, in which the time-honored memory of D-Day—particularly the vigilant leadership of Winston Churchill, Charles de Gaulle, and Franklin D. Roosevelt—was given a fulsome embrace. France may have helped the United States win at Yorktown but, Reagan insisted, America repaid the favor in spades 163 years later at Normandy. "One lesson of D-Day is as clear now as it was thirty-eight years ago," Reagan stated in his radio broadcast, which aired that day in the United States. "Only strength can deter tyranny and aggression." In taped remarks for French television Reagan echoed the same sentiment: "D-Day was a success, and the Allies had breached Hitler's seawall. They swept into Europe, liberating towns and cities and countrysides, until the Axis powers were finally crushed. We remember D-Day because the French, British, Canadians, and Americans fought shoulder to shoulder for democracy and freedom—and won."

The term "D-Day" was not coined for the Allied invasion. The same moniker was given to the attack date of nearly every offensive during World War II. It was first coined during World War I, before the massive United States attack at the Battle of Saint-Mihiel in France on September 12, 1918. The "D" was shorthand for "day." The expression literally meant "Day-day" and signified the day of an attack. By the end of World War II, however, the expression was synonymous with only one date: June 6, 1944. Evoking the fact that over 150,000 Allied troops had stormed the Normandy coast thirty-eight years before, at the cost of more than 10,500 dead, wounded, or missing men, Reagan pledged that America would do it again if it meant freeing France from a totalitarian aggressor like the Soviet Union. In the second year of his presidency, the resurrection of D-Day—the greatest amphibious landing in world history—as a defining moment in recent American history was already a central part of his anti-Soviet Cold War oeuvre. He closed his radio remarks by reading President Franklin D. Roosevelt's inspired D-Day prayer to the nation. According to historian Jon Meacham, author of *Franklin and Winston*, when FDR had read the prayer on the radio, he was playing the part of "national pastor"; now Reagan, in a far less dramatic historical moment, was doing the same.

Two years later, on June 6, 1984, Reagan came back to France for the fortieth anniversary of D-Day. The emotive speech he delivered at the windswept Normandy promontory looking out over the English Channel—known now in history as the Boys of Pointe du Hoc* address—was the opening salvo to a new American indebtedness to World War II veterans. By honoring the daring action of the 2nd Ranger Battalion—225 young Army volunteers whose mission was to climb the treacherous 100-foot-high Pointe du Hoc cliff while being shot at by entrenched German soldiers—he was paying tribute to an entire generation. (Out of those 225 "boys," only 99 survived the amphibious assault.) Later that afternoon the President also gave a stirring oration at Omaha Beach, not far from where the plaque on a bronze memorial statue honors "The Spirit of American Youth Rising from the Waves." By the 1980s, these youths were aging gray hairs, the new old. "Pointe du Hoc and Omaha Beach were Reagan's signature moments," Ken Duberstein, a former White House chief of staff, recalled. "If I have one enduring memory of Reagan, it's the way he crisply saluted World War II veterans that afternoon. These were his

*Sometimes—as in General Omar Bradley's *A Soldier's Story*—the Parisian French spelling Pointe du Hoe is used; I use the traditional Norman spelling throughout.

guys. Then there was the moving visual of Reagan walking with Nancy amongst all the grave sites, which looked like miles and miles of white crosses. That was an unforgettable moment of absolute reverence for the World War II vets. As president, Ronald Reagan delivered three unforgettable speeches: Pointe du Hoc, the *Challenger* disaster, and the Berlin-tear-down-this-wall number. But it was the first of these—Pointe du Hoc—that set the tone for the others."

Before Reagan dubbed the 2nd Ranger Battalion the Boys of Pointe du Hoc, they were known as Rudder's Rangers—named after their feisty, tough-minded commanding officer, Colonel James Earl Rudder. Historian Ronald L. Lane's *Rudder's Rangers,* first published in 1979, remains the indispensable book on the World War II years of this remarkable fighting man. Another invaluable book on the battalion is Robert W. Black's riveting *Rangers in World War II,* a definitive popular study of these elite fighting men. From an archival perspective, the Eisenhower Center for American Studies at the University of New Orleans, of which I served as director from 1994 to 2004, has numerous transcribed oral histories of various U.S. Army 2nd Battalion survivors. Never before have these oral histories been so fully tapped as in the writing of this book.

What distinguishes this deliberation from other D-Day narratives is the interweaving of the Battle of Pointe du Hoc (June 6, 1944) with the ascendance of President Ronald Reagan's New Patriotism (June 6, 1984).* With the assistance of

*The New Patriotism was Reagan's attempt to build a political consensus among Republicans, independents, and conservative Democrats based on an unflinching devotion to all things American.

the Ronald Reagan Presidential Library in Simi Valley, California, I've been able to document how and why our fortieth president played a seminal role in launching the great reappreciation of World War II veterans that swept over America in the 1980s and continues today largely unabated. If it hadn't been for Reagan's two elegiac June 6, 1984, homilies—written by Peggy Noonan (Pointe du Hoc) and Anthony Dolan (Omaha Beach)*—there may never have been Stephen Ambrose's *Band of Brothers,* Tom Brokaw's *The Greatest Generation,* Steven Spielberg's *Saving Private Ryan,* or numerous memorials—like the National D-Day Museum in New Orleans—built to exalt the citizen soldiers who liberated Europe. "Reagan's performance in Normandy demonstrated the timing, dramatic sense and attention to detail," biographer Lou Cannon wrote in a June 2004 obituary in the *Washington Post,* for "which the White House staff was famous during his presidency."

It was reporter Michael Dobbs of the *Post* who best put his finger on the 1980s-style New Patriotism, which was unleashed across America to commemorate D-Day plus forty years. "Forty years after D-Day, Normandy is bracing itself for a new invasion by a foreign army of 30,000 war veterans, journalists, television crews, tourists and secret service agents, with five monarchs, two presidents and a platoon of prime ministers in the vanguard," Dobbs forecasted on June 1, 1984. "For the politicians, notably President Reagan, the fortieth anniversary of the D-Day landings can be seen as a valuable public relations exercise at a time when election

*Both speeches are included in full in the appendix.

campaigns are in progress on both sides of the Atlantic. For the press and television, it represents one of the most colorful international news events since Britain's royal wedding three years ago. For veterans, most of them now well over sixty, the ceremonies could turn out to be the last opportunity for recapturing memories about the most momentous day in their lives. According to officials at the U.S. cemetery at Colleville-sur-Mer, a surprising number are taking advantage of retirement to return for the first time with their wives and families."

And return they did that June. Anywhere from 10,000 to 15,000 U.S. D-Day veterans made the transatlantic odyssey.* Their reasons for doing so were legion. Nearly all, however, wanted to pay obeisance to the past, to those Armed Forces members who had perished for freedom during the Second World War. The pre-anniversary D-Day hype orchestrated by the Reagan administration allowed many of these veterans, for the first time, to finally debrief their unknowing wives, children, and grandchildren about the grotesque horrors they had witnessed. No more buried remembrances. With Reagan as president, the time had come for the World War II generation to speak out, even if it was a psychologically gut-wrenching proposition.

By 1984 the stars were aligned for thousands of these stoic war heroes to finally offer eyewitness testimonials for posterity's sake. It was, in essence, a generational reckoning. Although these veterans had been allergic to boasting, President

*Peter Almond of the *Washington Times* estimated on June 5, 1984, that 30,000 D-Day veterans were returning to Normandy that month, 10,000 of them from the United States. Approximately 1.5 million men landed in Normandy on D-Day and the week that followed.

Reagan and the national media convinced them that telling "what happened" was their final duty to their country. Just as Supreme Allied Commander Dwight D. Eisenhower had ordered the U.S. Army to film the concentration camps immediately after their liberation by Allied forces so the Nazis would forever be held responsible for their sadistic war crimes, so too did these World War II veterans, forty years after D-Day, come forward with painful remembrances. It was as if they were finally ready, like the timeworn characters in Edgar Lee Masters's *Spoon River Anthology*, to walk onto center stage for a few moments, take a self-deprecating half bow, and recount their personal story of wartime survival to younger generations.

Sensing the zeitgeist—and in large part for political purposes—President Reagan became the Greatest Generation's self-appointed spokesperson on June 6, 1984. Although he never fought in World War II, Reagan had served in the Army Air Corps, eventually becoming a captain. All of the propaganda movies he made then were, in essence, a dress rehearsal for June 6, 1984. His dramatic words that day triggered the so-called Greatest Generation phenomenon, which swept across the United States largely unabated until the tragedy of September 11, 2001, brought a new group of heroes—particularly rescue workers, dedicated firemen, and city policemen—to the national forefront. In the early 1980s, with the impeccable timing of a maestro, Reagan galvanized the World War II generation into performing one last task: reminding a nation cynical after Vietnam and Watergate that America truly was still the shining city on the hill. What Reagan understood was that compared with the testimony of an Army Ranger who, climbing the Pointe du Hoc cliffs, had been forced to watch a buddy drown in the turbulent English Channel or a young officer get his legs blown off by a Nazi mine, 1970s slogans like "Acid, Amnesty and Abortion" were political throwaway lines of a decadent and largely self-indulgent recent past.

What President Reagan was trying to engineer in 1984—using the World War II generation and the American flag as his platform—was the creation of a combustible patriotism, one that would spread like wildfire: an extension of his 1980 presidential campaign's embrace of increased military spending and upgrading the Armed Forces. He essentially wanted to turn the clock back to an unambiguous black-and-white era when, as Stephen Ambrose opined in *Citizen Soldiers,* the sight of a GI meant joyous cheers from those communities occupied by Fascist troops. "The 'we' generation of World War II (as in 'We are all in this together') was a special breed of men and women who did great things for America and the world," Ambrose wrote. "When the GIs sailed for Europe, they were coming to the continent not as conquerors but liberators."

From Reagan's perspective, too many young people knew about the horrendous atrocities at My Lai and not enough about the raw gallantry of D-Day. Although he delivered numerous triumphalist paeans during his 1980 presidential campaign and was constantly challenging the Soviet Union to free its own people from Communist tyranny, it was his vivid June 6, 1984, Boys of Pointe du Hoc speech, carried live on major U.S. television networks, that raised the bar on an entire generation poised on the cusp of collecting Social Security checks. Even Reagan's sharpest liberal critics—like ABC's

White House correspondent Sam Donaldson—had to admit that the President had delivered that Normandy speech with an unmistakable inner conviction, one that started the transformation of his historical image from "conservative president" to "America's president." Journalist Michael Barone of *U.S. News and World Report,* in his fine book *Our Country: The Shaping of America from Roosevelt to Reagan,* believed that based on the potency of Reagan's Pointe du Hoc devotional "the election was over by June 6."

Up until that speech, the World War II event most Americans honored was Pearl Harbor Day (December 7). After that, August 6, the day the United States dropped a nuclear bomb on Hiroshima, was probably the best remembered anniversary moment of the war. The problem, from the indefatigable Reagan's optimistic point of view, was that *we lost* the Battle of Pearl Harbor. How do you create a New Patriotism over American unpreparedness and battleships on fire? President Franklin D. Roosevelt was effectively able to use Pearl Harbor as a rallying cry back in December 1941, a clarion call that bespoke determined revenge. But there were ambiguities surrounding Pearl Harbor, an event anti-FDR isolationists deemed "the Great Deception." Blue-ribbon investigative commissions—nine of them, to be exact*—were created, for example, to discuss what went

*The Knox investigation (December 9–14, 1941); the Roberts commission (December 18, 1941–January 23, 1942); the Hart investigation (February 12–June 15, 1944); the Army Pearl Harbor Board (July 20–October 20, 1944); the Navy court of inquiry (July 24–October 19, 1944); the Clarke investigation (August 4–September 20, 1944); the Clausen investigation (January 24–September 12, 1945); the Hewitt inquiry (May 12–July 11, 1945); and the Joint Congressional Committee (November 15, 1945–May 23, 1946).

wrong in Hawaii and which officer was responsible for the shocking debacle. As for Hiroshima, many people around the world—and peace activists at home—thought it stood for unconscionable genocide. While most Americans surveyed since 1945 believed that President Harry S. Truman was more than justified in dropping an atomic bomb on the city of 350,000, for the simple reason that it eliminated the need for an invasion of the Japanese home islands, it wasn't a military event that conjured up unrestrained, full-bore, patriotic pride. When discussing Hiroshima, people often turned defensive, saying things like "We had to drop it" or "What choice did Truman have?" The wholesale destruction of a city—where civilians were killed in unacceptably high numbers—was hardly the best symbolic event on which to build a New Patriotism movement.

But D-Day? The liberation of Europe? The Boys of Pointe du Hoc? That was a different, "perfect cause" story to sell. There were no unanswered questions or moral ambiguities about the Normandy invasion: our young martyrs liberated Europe from the macabre stranglehold of the diabolical Hitler. Period. American GIs gave their lives so others could live free. D-Day was clearly America's finest hour—and Reagan knew it. The real question about D-Day was how to boil it down to a condensed *Reader's Digest*–like feature where American soldiers, not the celebrated armada itself, were the emblematic heroes. Most Americans know very little about George Washington's Revolutionary War battles at Brandywine or Long Island. But we all know he froze—along with his men—at Valley Forge. Why? Because that ordeal exudes inherent human drama: men suffering from frostbite, dysentery, pestilence, and scurvy because they believed in democracy and liberty. By the time of his June 6, 1984, speech, the cliffs at Pointe du Hoc had become to Reagan an enduring, symbolic American locale just like Valley Forge, even though it was situated on French soil.

NBC News anchor Tom Brokaw—who had been in Normandy filming a documentary—had not yet coined the phrase "the Greatest Generation" when Reagan delivered his fortieth-anniversary orations. That would come a decade later, in 1994. But Reagan, as a charter member of that generation, had intuitively believed for over four decades that the Americans who died at Normandy and other such sacred battlefields truly were cut from a special cloth. Along with the savage depravations of the Great Depression, the Second World War had been the definitive event of Reagan's generation. Those who had survived the global cataclysm—like himself—still shared the old-fashioned sense of duty, honor, and country that the war instilled. Ever since Pearl Harbor—and long after V-J Day—the veterans of World War II, who included Reagan in their ranks, never wavered from their determined motto: "We're All in This Together."

Anybody who was ever close to Reagan tells stories—his stories—about how much he admired the American fortitude displayed during World War II at such faraway places as Midway, Iwo Jima, Okinawa, Sicily, Anzio, and, of course, Normandy. "Reagan was a master storyteller," Brokaw recalled in 2004. "When you go back over his eight-year presidency, like we did when he died, it's the way he brought real people into his speeches that stands out. I believe that Reagan's the Boys of Pointe du Hoc speech, which Peggy

Noonan wrote, was the beginning of the rekindled awareness of what we owe the World War II generation. He opened the window, so to speak, so we could all see through it. . . . Generally speaking, Omaha Beach is too complicated for people to fully grasp. There was just so much going on. But, as Reagan recognized, you could get your hands around Pointe du Hoc. The standoff there was about individual valor, Rangers climbing a cliff under fire, yet they kept on going. But nobody made much of a fuss over D-Day before Reagan. It was the perfect military moment for him to honor, and he seized upon it."

Commandos' monument in Scotland. (Courtesy of JoAnna McDonald)

1

DARBY'S RANGERS

As a movie actor, history buff, and unabashed nationalist, President Ronald Reagan, while he prepared for the fortieth anniversary of D-Day, was keenly aware of the sheer power the word "ranger" conjured in the American imagination, even if his facts about their early combat antics were—as historian Garry Wills claims—often of the Disneyland triumphalist variety. Ever since Jamestown was established by English settlers in 1607, "rainger" or "ranger" had become part of the New World vocabulary. Feeling vulnerable to attack by Native Americans, early colonists dispatched armed scouts to roam the wilderness and, if necessary, exterminate potential enemies before any Indians could wreak havoc on their Christian communities. The daily reports of these scouts often said something like "ranged twenty miles yesterday" or "too rainy to range far," so it didn't take long for the term "ranger" to stick. They were, in effect, wilderness patrolmen. They realized early on that European-style warfare was unsuitable to America's roiled terrain. Rangers, adjusting to topography, adopted the stealth tactics and nomadic ways of the various Indian tribes who freely roamed the eastern seaboard, and they aroused the hostility of the Native Americans inhabiting it. Such combat was almost the antithesis of European warfare, which at the time consisted of much maneuvering and very little fighting. "In the last decades of the seventeenth century rangers acting as quasi-military units first appeared in Massachusetts and Virginia," historian Jerome Haggerty noted. "At that time combat in North America was influenced by the wilderness."

As soon as the French and Indian War (1756–1763) broke out, a Natty Bumppo–like New Hampshire backwoodsman named Robert Rogers, livid that Native American warriors constantly ambushed local forts and homesteads in hit-and-run raids, sought revenge. He signed up with the British Army to be a scout. Before long he received a commission as captain and organized a "ranger" team of a few dozen men trained for hand-to-hand combat against both the French and Indians. Officially they were the Ranger Company of the New Hampshire Provincial Regiment; a year later they had become His Majesty's Independent Company of American Rangers. Clad in distinctive green outfits, they eventually became known as Rogers' Rangers and soon attacked French fortifications along Lake Champlain and Lake George.

Accordingly, Rogers was dubbed Wabi Madaondo, or white devil, by the Indians who crossed his path. He, in turn, said that Native Americans had "revengeful" dispositions in his 1765 book, *A Concise Account of North America.* "Throughout the French and Indian War, Rogers, and his Rangers, continued to wage unconventional warfare throughout the upper New York, lower Canadian, and even French West Indies regions," historian J. D. Lock wrote. "Following the war, there were periodical 'revivals' of Rogers' Rangers in support of British Operations until 1763 when they were 'paid off' for the first time."

The very notion of these rangers—often with smeared war paint on their faces and waving sharp knives—caused unmitigated fear among the various Native American tribes (particularly the Abenaki St. Francis Indians) warring against the British empire. Teaching his mobile recruits how to live off the land, Rogers often took advantage of subzero winter weather to catch his slumbering enemies off guard. By using specially made snowshoes, Rogers' Rangers could cross frozen rivers and lakes with bold ease, startling their unexpecting adversaries. Unusual for military men of the colonial era, these hardened rangers sought to unnerve their opponents by engaging in everything from blood-curdling yells to midnight attacks. Lightning-fast raids were their specialty. Decimating the enemy was all that mattered. Almost intuitively, Rogers' Rangers understood what, according to historian Robert W. Black, would become the modern-day mantra for the U.S. Army Rangers: "It is all in the heart and the mind."

During the American Revolution, Rogers volunteered to serve George Washington; the general refused his help, fearful he was a loyalist spy. Snubbed, Rogers abandoned the call for independence and instead organized a battalion of pro-British loyalist commandos known best as the Queen's Rangers. If Washington didn't trust him he would remain loyal to King George III. Back in Virginia, however, a new volunteer outfit, clad in coonskin caps and equipped with long rifles and hunting knives, took the fight to the Redcoats. Led by Captain Daniel Morgan, this ranger battalion did whatever chore was needed, from toggling across the swift currents of New York rivers to doing surveillance scouting in the pristine Virginia countryside. A great motivator of men, Morgan used a recruitment test that became legendary among Washington's Continental Army. He printed up numerous broadside illustrations of King George's robust head. To join his rangers you had to be able to shoot the British monarch in the face from 100 yards away, usually on the first try. Because of this hateful recruitment practice, Morgan was deemed a "war criminal" in London.

But it was the lore surrounding the Texas Rangers that President Reagan was most familiar with. In 1823, Texas was still controlled by Mexico. Enter Stephen F. Austin, the so-called Father of Texas, whom William Clark (of the Lewis and Clark expedition) anointed with "special trust and confidence" to become a new type of American frontier hero. With calm fortitude Austin organized Ranger units whose sole mission was to free Texas. With Colt six-shooters on their hips and Kentucky rifles flung over their backs, they defended towns like San Antonio, Laredo, and Brownsville. Before long their notorious reputation for gung ho fearlessness had grown into the stuff of legend. Texas Rangers emerged as role models for a new type of indomitable American fighting man.

During the Mexican-American War of 1846–1848, the Texas Rangers once again did their part, helping Brigadier General Zachary Taylor control the Rio Grande. Meanwhile, in the northern states, West Point was still considered the proper place to prepare a young man for the vicissitudes of combat, not the wild cactus country along the Brazos River, which ran 840 miles across Texas to its mouth on the Gulf of Mexico. But throughout the more agrarian South the idea of a Texas Ranger perched high atop a sleek stallion, his finger always poised on his trigger, spouting out words like "freedom," "independence," and "liberty," had become part of the enduring folklore of the region. Studying Aristotle or Thucydides along the Hudson River seemed effete compared with killing Apaches or Mexicans along the lawless border. Captain Jack Hays, a colonel in the 1st Regiment, Texas Mounted Volunteers, became known for a prayer he delivered before the Battles of Monterrey and Mexico City: "O Lord, we are about to join battle with vastly superior numbers of enemy, and, Heavenly Father, we would like you to be on our side and help us; but if you can't do it, for Christ's sake don't go over to them; but lay low, keep dark, and you'll see the damndest fight you ever saw in all your born days." A hundred and fifty years later, Hollywood actors like Chuck Norris played rough-and-ready Hays-like Texas Rangers on television, macho action stars, white hats fighting evil wherever it reared its ugly head.

Given the Texas Rangers' lore, it's not too surprising

that when the cannons went off at Fort Sumter, South Carolina, and the Civil War commenced, the Confederate Army started organizing Ranger-style units. Between 1861 and 1865, in fact, about 400 Ranger outfits were created. By far the most infamous was Mosby's Rangers. Commanded by John S. Mosby of Virginia, who was nicknamed the Gray Ghost throughout Appalachia, the unit brought the fight to the Union Army like true guerrilla warriors. They blew up train depots, intercepted Yankee couriers, burned crops, sabotaged Union base camps—and did it all with a haunting Rebel yell. By 1865 Mosby had risen to colonel, cited for meritorious service by Robert E. Lee—more so, in fact, than any other Confederate officer. "His exploits are not surpassed in daring and enterprise by those of *petite guerre* [guerrilla or small warfare] in any age," wrote Major General J. E. B. Stuart of Mosby in February 1864. "Unswerving devotion to duty, self-abnegation, and unflinching courage, with a quick perception and appreciation of the opportunity, are the characteristics of this officer."

Given such an inspiring history, it is somewhat surprising to realize that the U.S. Army did not organize Ranger units until 1942. (Part of the reason, one suspects, is that they had become synonymous with the Confederacy.) The impetus of the Rangers' rebirth was the so-called commandos, whom British prime minister Winston Churchill started employing in 1940 after the Nazi blitzkrieg overran Europe. Once the British fled Dunkirk across the moatlike English

Channel, a sense of desperation had struck Downing Street. Could the United Kingdom survive the German onslaught if it only took a defensive posture? Shouldn't Britain find new ways to take the fight to the enemy? Wasn't an elite fighting force necessary? Churchill instructed the Ministry of Defence to increase the importance of British commandos who could conduct secret amphibious operations, destroy Nazi targets, collect intelligence, and then covertly return home across the Channel.

From Churchill's point of view, since Germany had had such success with their elite units known as storm troopers, he would now one-up the Nazis with his commandos. In order to liberate Europe, he believed, "striking companies"—able to engage in both covert and overt actions—would have to be formed at once. Writing to his chief of staff, Lord Ismay, on June 6, 1940, Churchill demanded an elite new group of what he deemed the "hunter class" who would terrorize the various Nazi fortifications erected along the coasts of Europe and North Africa. Refusing to mince words, he wanted these "commandos" to excel in the art of "butcher and bolt." Churchill's men would be trained in demolition, destruction, and assassination. They would be, in essence, self-contained killing machines. He envisioned the commandos as "a steel hand from the sea."

There is some debate as to when and where the term "commando" was first used. It may have been adopted during the French and Indian War, when General John Forbes wrote to Colonel Henry Bouquet during the campaign against Fort Duquesne (modern-day Pittsburgh) in 1758.

Others attribute the origin to Lieutenant Colonel D. W. Clarke of the British Army, an early-twentieth-century expert on the history of mobile raiding units used in various guerrilla wars then raging from the Middle East to South Africa and the Philippines. As it turns out, Clarke knew firsthand the absolute success of Dutch "commandos"* used during the Boer War in South Africa (1899–1902). These Boer settlers instituted merciless scorched-earth tactics against their foes. Churchill knew this very well, since he had served as a war correspondent for London's *Morning Post* in South Africa, where he was even taken prisoner. "In selecting the name Commando, Clarke must have been aware that Winston Churchill had gained national fame as an escaped prisoner during the Boer War," Robert W. Black surmised in *Rangers in World War II*. "The name Commando was certain to touch memories of youth and glory in the prime minister."

Back in the United States, General Dwight Eisenhower, assistant chief of the Army War Plans Division, was encouraged by Britain's sped-up efforts to train self-contained fighting units who knew no fear. But as a staunch believer in traditional warfare, Eisenhower was more concerned about building up large infantry divisions brimming with modern tanks and M3 half-tracks that could overwhelm the Third Reich. Nonetheless, Eisenhower was also an open-minded pragmatist. Within the U.S. Army a brilliant colonel from Texas, Lucian K. Truscott, was pushing General George Marshall to start up its own commando training program. On June

*The Dutch Boer term was "kommando," for troops of any size.

1, 1942, Marshall ordered the creation of an "American commando" unit. To regular Army traditionalists like Eisenhower and Marshall, adopting such SWAT team approaches to warfare could possibly prove perilous to troop morale. The mere fact that special forces—made up of volunteers—were suddenly needed inferred that the regular ones were not up to the job. In addition, selecting the crème de la crème out of each company would drain existing divisions of their best leaders. Truscott, however, eventually got the go-ahead. He went to Carrickfergus, Northern Ireland, to study British commando training and report back what he learned. His transatlantic mission was attached with a lukewarm endorsement from Eisenhower: "If you do find it necessary to organize such units," Eisenhower instructed Truscott, "I hope you will find some other name than 'commando,' for the glamour of that name will always remain—and properly so—British."

No country, it seems, enjoys macho male fighters and iconoclastic daredevils quite as vociferously as the United States, the land of John Wayne and Rambo. After all, its capital city was named after George Washington, the world-renowned guerrilla fighter of the Revolutionary War. The "Father of the Navy," John Paul Jones, continues to live on in the national folklore as the gutsy captain who uttered "I have not yet begun to fight." Hardscrabble backwoodsmen like Davy Crockett and Andrew Jackson became the post–Revolutionary War generation's first heroes. Given this penchant for frontier cunning and true grit, it's not surprising that the American public excitedly embraced the whole concept of British commandos liberating Europe from Nazi totalitarianism. As historian

1st Ranger Battalion in basic training mode. (Courtesy of JoAnna McDonald)

JoAnna McDonald noted in her book *The Liberation of Pointe du Hoc,* even the *New York Times* enthused that the commandos were "super-guerillas, the modern Apaches of the British armies." An article in the venerable *Harper's Weekly* by Bruce Thomas (which McDonald also references) took the commando enthusiasm a step further, reaching true comic-book-like characterizations in his lengthy article. According to Thomas, the British commando was a combination of Peck's Bad Boy, Robin Hood, Tarzan, Superman, Daniel Boone, and a Boy Scout. "He is the cagiest and canniest fighting man on the loose to-day," Thomas wrote, as if coming straight from the pages of an 1840s *Crockett Almanac.* "His average age is twenty-seven. . . . [They are] the world's number one guerilla fighters and unconventional scouts."

It was left to Colonel Truscott to come up with an appropriate name for a new "American commando" outfit.

Luckily for him an obvious candidate was already in the air. Novelist Kenneth Roberts had recently published *Northwest Passage* (1937), about the exploits of rangers during the French and Indian War. A movie version starring Spencer Tracy and Walter Brennan was released in 1940, garnering strong box office receipts. The popularity of Roberts's "Indian fighter" novel/movie, plus the fact that Truscott was a Texan, well steeped in the Lone Star lore of Austin and Hays, naturally led to "ranger" surfacing to the top of his name list. In straightforward fashion, Truscott recalls his decision in his 1954 memoir, *Command Missions: A Personal Story*. "Many names were recommended," he wrote. "I selected 'Rangers' because few words have a more glamorous connotation in American military history. In colonial days, men so designated had mastered the art of Indian warfare and were guardians of the frontier."

Having a moniker and a mission, however, was the easy preliminary step in creating a special forces outfit. The key to building a first-rate military is always based on first-rate training. Finding the right leader for the 1st Ranger Battalion would be of the foremost importance. The United States—and the world—is lucky that Major William O. Darby of Fort Smith, Arkansas (West Point class of '33), was available for the task. Honest and tough as nails, Darby, like a Lombardi-style football coach, knew how to train young men. As a field artillery officer serving as an aide to the U.S. Army in Northern Ireland, he inspired loyalty, respect, and, at appropriate times, fear in his eager recruits. His greatest gift was in knocking the "hot dog" factor out of young men overly impressed with their athletic prowess or all-night stamina. With America now at war against Germany, Italy, and Japan, Darby was looking for approximately 35 officers and 450 men who were highly intelligent, morally fit, physically healthy, and psychologically sound. Darby, in sum, was looking for the best of the best—whiners or wimps of any kind need not apply. He found the soldiers he was looking for. They came from every state in the union. They were all volunteers. They all wanted to destroy Hitler and Mussolini and Tojo.

More than 2,000 men applied for the honor of becoming a Ranger. The 700 who survived the first round of cuts came seemingly straight out of a populist Carl Sandburg poem. As JoAnna McDonald points out, the youngest was Private Lemuel Harris, only eighteen, from Pocahontas, Virginia. The oldest was thirty-five-year-old Sergeant J. B. Commer of Amarillo, Texas. Many of these applicants' occupations spoke directly of the regions they hailed from: bullfighter, cowboy, oysterman, burlesque operator, church deacon, oil speculator, railroad worker, coal miner, tomato farmer, barber, insurance salesman—just reading the roll call of names conjured up an image akin to a massive Benton mural. Before long, after a few weeks of tough training, 180 men were dropped from the competitive program. They were good men, made of the finest American fiber, but not quite good enough to become Rangers. "The outfit that can slip up on the enemy and stun him with shock and surprise," Darby declared, "that is the outfit that will win battles, and that is the outfit I want."

On June 20, the remaining 520 were shipped off to the British Commando Training Depot at Achnacarry Castle,

Scotland, isolated among seals and otters far north in the lush rolling hills of the windswept Highlands along the River Ness. U.S. Army advance scouts had reported back to Truscott that a key element of the commandos' swagger was that they *believed* they were the finest fighting men anywhere. Pessimism, for the most part, was not in their makeup. Optimism—and patriotic platitudes like "Saving the World for Democracy" and "Freedom from Fear"—coursed through their veins. "Their *esprit de corps* must be kept at its peak by frequent allusions to their superiority to troops trained in ordinary ways," Lieutenant Alfred H. Nelson wrote to Truscott. He added a competitive jingoistic aside: "We do believe that it is possible by use of some of the British methods, plus some that I am sure will be devised by our own commanders, to train American personnel to do a better job in any situation than that done by their British forerunners."

Over the decades Hollywood has made numerous films showing just how hard the U.S. Army trains its elite troops, but the most accurate description available of what it was like to be a recruit in the 1st Ranger Battalion won't be found in an Oliver Stone or Taylor Hackford movie. It can, however, be gleaned from William Darby's 1980 memoir, which is appropriately titled *Darby's Rangers: We Led the Way,* written with the assistance of William H. Baumer. "We marched swiftly, swam rivers or crossed them on bridges made of toggle ropes," Darby reports. "There were cliffs to climb, slides to tumble down, and when all that was quite enough we played hard games." Oral histories of 1st Ranger Battalion veterans who served in Great Britain tell of log-roll competitions, boxing matches, speed marches, jujitsu bouts, bayonet drills, swamp treks, starvation tactics, mock battles, and push-up marathons. Salutes were judged on technique, and uniforms were somehow supposed to be clean after drilling in a bog. When the whistle blew you responded as if your life depended on it. This was basically "think fast or die" training. The paradoxical trick was to both follow orders and think for oneself. According to Ronald L. Lane, in *Rudder's Rangers,* this first wave of Rangers had to scrap notions of "sportsmanship" to learn that war was a dirty business without set rules. They had to become, in effect, "versatile in the art of killing."

And then there was the weapons training. They were instructed to use the M1 carbine, M1 Garand rifle, Thompson submachine gun, Browning automatic rifle, and 60-mm M2 mortar. Five different grenades were passed out, ranging from the basic offensive hand grenade to a thermite grenade and one that merely illuminated the sky. Because friendly fire—and accidents—claimed so many lives in combat, the Rangers were taught how not to make mistakes with weaponry that was meant solely for the enemy.

What is most relevant to this study, however, is the way this first class of U.S. Army Rangers was trained to climb towering cliffs by using spikes and rope ladders. Various fourteen- to twenty-foot walls were constructed in open fields. The Rangers were taught to scramble up them, quickly, then once on top, leap down into the mud, roll over, and, with an automatic weapon in hand, start blasting away at various movable targets. Sprinting up small mountainsides while mortar shells exploded all around them was another part of their grueling training. Even for unflappable young men, trudging up a hill weighed down with equipment, with explosions going off all around, was unnerving. Eager youths were being molded into hard-shell men.

Those Rangers who graduated from Achnacarry on August 1, 1942, were soon relocated to Argyll for a month of rigorous amphibious training with the British Royal Navy. Lord Louis Mountbatten, who had been appointed head of Combined Operations by Churchill, instructed Truscott to select around fifty of his best Rangers to help Canada and Britain in an upcoming, top secret raid against the so-called Atlantic Wall, which the Nazis had fortified along the entire coastline of France. As it turned out, the first Americans to die in a World War II battle in Europe, in fact, would be members of the 1st Ranger Battalion when they raided the Normandy town of Dieppe.

The high-risk assault on the German-occupied port of Dieppe began early on August 19. An armada carrying almost 250 Allied vessels left England, hauling more than 6,000 troops and sixty tanks. The Allied forces immediately took heavy casualties. With Allied air squadrons bombing Nazi fortifications along the Atlantic coast, the brave Canadian commandos, along with their American and British counterparts, tried to scale thirty- to hundred-foot-tall cliffs, overtake the enemy, and then march into the French resort. But from the start, nothing worked out as planned. Some commandos and Rangers, however, actually made it over the imposing cliffs. One Ranger, Corporal Franklin Koons, in fact, successfully killed the first Nazi he laid eyes on. The Americans had arrived. And there was other good news. For the most part the big German guns guarding the Norman town were annihilated. Seeing their smoldering rubble was a good omen for the Allies' future.

But the immediate military reality was inescapable: British, Canadians, and Americans were driven back into the sea. By any traditional standard, the cross-Channel raid was an unmitigated disaster. "It is a small wonder that the very word *Dieppe*," historians William Whitehead and Terrence McCartney Filgate wrote, "chills the hearts of an entire generation." Of the 6,100 British and Canadian troops who took part, over 3,500 were either killed, captured, or deemed missing in action. Some men drowned trying to get to the beach, while others fell to their death attempting to climb the uncooperative cliffs. Not a single tank managed to get into the town. A staggering 160 Allied planes were shot down, while the German Luftwaffe lost only 48. Stung by the setback, a weary Churchill was, for intelligence-gathering reasons, sickened to learn that the Nazis had captured over 1,000 of his men. As for the 50 specially chosen Rangers, 6 were killed, while another 7 were wounded in action and 4 captured by the enemy.

In the wake of Dieppe, the 1st Ranger Battalion was ordered to move to the town of Dundee for an intensified program in coastal raiding. "The Rangers practiced attacking pillboxes, gun batteries, and other coastal defenses," historian Mir Bahmanyar wrote in *Darby's Rangers, 1942–45*. "In Dundee the Rangers stayed with families in town as there were no barracks available to them."

While Dieppe was only a raid, Hitler's propaganda chief, Dr. Joseph Goebbels, immediately marketed it for global consumption as a full-fledged invasion. Berlin newspaper headlines—under Goebbels's orders—reveled in the result: "Catastrophic Defeat a Setback to Invasion" and "What Does Stalin Say About This Disaster to Churchill's Invasion?" If one wanted to render a tough historical verdict, it was as if Churchill had another Gallipoli-like debacle on his hands. "You've got to start one day fighting the enemy," Lord Lovat of No. 4 Commando Company had warned Churchill, "but you don't start by landing frontally in daylight at a defended port like Dieppe."

With the benefit of hindsight, however, Dieppe looks a little less like an utter Allied failure than it seemed at the time. Churchill himself thought the raid was mostly successful. "Dieppe occupied a place of its own in the story of war, and the grim casualty figures must not class it as a failure," he wrote. "It was a costly but not unfruitful reconnaissance in force." Without question, the amount of Allied casualties incurred vis-à-vis the Nazis was unacceptable. But this D-Day dress rehearsal had positive side effects. From Dieppe onward, U.S., British, and Canadian troops would work in closer unison. For example, the 1st Ranger Battalion cooperated more closely with a group of No. 1 and 6 Commandos known as "Commaradies." Both commandos and Rangers would intensify their efforts to practice amphibious landings, climbing cliffs, destroying pillboxes, and liberating coastal towns. Meanwhile, the American media, gleeful that the Rangers had participated in the large-scale raid, bannered its own more upbeat headlines, one reading "U.S. and Britain Invade France."

Darby's Rangers—like Rogers' Rangers of old—were

taking on a mythic stature. By October they were given Operation Torch (the name for the invasion of North Africa) assignments to take out a two-gun battery perched on a jutting cliff in the harbor of Arzew, Algeria, which lorded over the primary landing beaches. On the evening of November 8, they succeeded. For the next couple of months, as they were assigned to an Army Training Center at Arzew, the Rangers stayed in Algeria or next-door Tunisia. Now that the Rangers "owned" Arzew, they were responsible for guarding the hospitals, ammo depots, and town center. After that they secured Gafsa and El Guettar and kept on going. If the North African campaign was successful, sacking Sicily next would be a distinct possibility. Then, in late January 1943, they were truly unleashed.

By February 11, they had successfully devastated Italian forces guarding Sened Pass. "Sened Station was a German supply station for ammunitions and was kept heavily guarded," Ranger Bing Evans recalled. "They also had troops in there ready to move up to the front."

Whether it was bridge demolition or river foraging or slitting throats of an Italian in broad daylight the 1st Rangers were, without question, the gutsiest new Army outfit around. They became pros at disembarking, largely undetected, from landing craft to conduct speed raids. "It's a prickly feeling you get on an invasion," Ranger Thomas Sullivan wrote his brother in August 1943. "We could see the big searchlights on shore and the sea was so rough I got very seasick—first time—we came in small assault boats. I was very tense and kept gripping my rifle. As we were about to land I was raring to go—about that time a wave dusted over and cooled me off.

We jumped into the water and ran ashore. Firing started and from there it was *c'est la guerre*." As Darby recalled in his memoirs, the Rangers of early 1943 had become a "miniature Stonewall Jackson force, operating like the Confederate foot cavalry in the Shenandoah Valley during our Civil War." Historian Ronald Lane put it more succinctly: the 1st Rangers in North Africa "literally slaughtered the enemy."

Perhaps the highest compliment in warfare occurs when your dire enemy nicknames you out of begrudging admiration. One doesn't need to know all the daily maneuvers of Field Marshal Erwin Rommel in North Africa to comprehend that his moniker "the Desert Fox" is a tribute to his cunning. So when the 1st Ranger Battalion annihilated the elite Italian Bersaglieri units at Sened Station, with Darby winning the Silver Medal for valor, an ominous nickname emerged for the outfit. The Italians now called the U.S. Army Rangers "Black Death." They were spearheaded with the task of invading Sicily and Italy. And praise came pouring in from Washington and London. While reconvening at Gafsa, a proud Darby corralled all of his men together for an impromptu talk. Eisenhower, impressed by the successes of the 1st Battalion, had claimed he was now a convert, completely "sold" on the Rangers.

Darby's Rangers would land on the left flank at Salerno in September 1943; still later, they dramatically seized the town of Maiori and the imposing heights of the Chiunzi Pass overlooking Naples and denied the pass to the Germans to counterattack the Allied beaches. They also looked down on the Germans in Naples and kept them under fire.

And the commendation spread like wildfire throughout

the ranks of the U.S. Army and beyond. By March 1943, with combat experience under their belts, a specially selected group of Rangers was sent back to the United States to help train the 2nd Ranger Battalion, which was being organized in a bustling Tennessee outpost called Camp Forrest.

Pointe du Hoc, France. (Courtesy of JoAnna McDonald)

2

RUDDER'S RANGERS

It was April Fools' Day 1943 when the United States activated the 2nd Ranger Battalion. However, this time around, training in Northern Ireland or Scotland was not part of the plan. Camp Forrest, one of the Army's busiest posts during World War II, located near Tullahoma, Tennessee, would be their initial training ground. Named after the brash Confederate cavalry general Nathan Bedford Forrest, the remote camp was designed to train infantry, artillery, engineer, and signal units. It was an 85,000-acre site erected 1,070 feet above sea level, the highest point between Nashville and Chattanooga. This vital facility became a frenetic beehive of military activity during the war years; Major

General George Patton, for example, even brought his 2nd Armored "Hell on Wheels" Division up from Fort Benning, Georgia, to participate in maneuvers. A little concrete was spilled, and guard towers erected, to give the Tennessee mudflat the illusion of permanence. Military personnel flooded the boomtown area. As Tullahoma grew, a so-called tent city was erected for the Rangers, since lodging quarters came at a premium. At the time of Pearl Harbor the population of Tullahoma was 4,500; by the time the atomic bomb was detonated on Hiroshima, it had grown to 75,000.

A few of the new arrivals were 1st Ranger Battalion fighters who came to Tullahoma ready to share eye-opening stories about the North Africa campaign, and to instill in the newly formed 2nd Battalion the fighting spirit of the 1st. But for them to succeed, it was imperative that the Army find the right commanding officer. A revolving door of potential candidates came and went, some lasting only a couple of days. It was tricky finding the right man, with the right temperament, to lead such an elite new fighting unit. Comprised of 27 officers and 404 enlisted men, the 2nd Battalion would have a company headquarters and six Ranger infantry companies. Every hour, it seemed, some high-ranking desk officer was reconfiguring the Rangers—they were a work in progress. A majority of the volunteers came from New York, Pennsylvania, and New Jersey. But Texas provided the hardscrabble man eventually tapped to be the commanding officer. In late June Major James E. Rudder was assigned command of the battalion. As historian Gordon L. Rottman explained, with the arrival of Rudder "things began to tighten up."

Known by his friends as Big Jim, the stocky, tough-talking, no-nonsense Rudder was born on May 6, 1910, in the west Texas town of Eden. He was one of six children whose father ran a livestock business near the Concho River. When he was seventeen, he was recruited to play football at Tarleton, a junior college in nearby Stephenville. He was a center. Determined to be an "Aggie," Rudder, who dug ditches to help his family survive the Great Depression, transferred to Texas A&M in 1930. Two years later, when the six-foot, 220-pound Rudder graduated, he had been commissioned as a second lieutenant of infantry in the Army reserve. Blessed with the chiseled Hollywood good looks of Glenn Ford, accentuated with a touch of Gene Hackman gruffness, Rudder was unabashed in his zeal for hard-tackling, take-no-prisoners, Texas-style football. In college, it was said, he loved to block more than breathe. And he was a gridiron leader. For a few years, before the war, he coached the Brady and Tarleton high school teams to winning records.

Rudder was called to active duty following Pearl Harbor. In rapid succession he attended Infantry School at Fort Benning, served as battalion executive officer at Camp Atterbury, Indiana, and went to the Army Command and General Staff College at Fort Leavenworth, Kansas. After bouncing around to various stateside posts, he was dispatched to Tennessee to make men out of boys. All of the eighteen- or nineteen-year-old volunteers who made up most of the 2nd Ranger Battalion recall the day Rudder arrived at Camp Forrest. He told his Rangers, "I'm going to work you harder than you've ever worked. In a shorter time than you can imagine, you're going to be the best fighting unit

The indomitable Lieutenant Colonel James Earl Rudder. (U.S. Army Military History Institute)

in this man's army." Drills became tougher. Double time became the mantra. Twelve-mile hikes turned to twenty-four-mile endurance treks. Catching one's breath was a "dilly-dally," guaranteed to cost you fifty additional push-ups on the spot. Obstacle courses sprung up in unexpected places, like unwelcome mushrooms in a field of green. If you had been bench-pressing 100 pounds, you were told to up the weight to 150 or 200 pounds. Boxing matches—with or without gloves—were encouraged. But Rudder cared deeply about each of his men. "He talked to you softly but firmly, like a big brother," one of his star Rangers, Leonard Lomell, recalled. "He inspired you to do your best. You just wanted to die for him."

Yet, in a number of important ways, basic creature comforts improved. There was an egalitarian fairness about the way Rudder treated enlisted men that commanded almost instant loyalty. He resourcefully wangled wooden barracks for his men, getting them out of the tent city. Instead of the daily Army C rations of corned beef hash, boiled carrots, canned peaches, and stale bread, soon-to-be Rangers were trained to be cooks. Indoor latrines were opened. Mosquito netting was provided in the barracks. And, most important, this dizzying six weeks of brutal training, the kind that would tax any man both mentally and physically, suddenly seemed to have a galvanizing purpose. Forget Darby's Rangers. They were Rudder's Rangers, and they were going to be *the* best. These men President Reagan would dub the Boys of Pointe du Hoc in 1984 started gelling as a first-rate battalion in the sticky summer days of Tullahoma, Tennessee. "To get into the Rangers, you had to volunteer," Ranger Ralph Goranson recalled. "At any time during training, you could volunteer out, because of the physical requirements, mental requirements, or just the fact that you just plain didn't fit. We had a high turnover at the beginning."

If Rudder had a particular genius, it was in eyeing talent. He was not impressed by mere brawn. A Ranger, he believed, had to have moxie and heart. Take, for example, the story of twenty-two-year-old Private William Petty of Cohutta, Georgia. A bright, pale-skinned former scholarship student at the University of Georgia, he had arrived at Camp Forrest as a private in the 30th Division of the Tennessee National Guard. Unfortunately, when parachuting one afternoon, he had broken both legs. Everybody noticed Private Petty had a terrible limp; he waddled when walking, resembling Charlie Chaplin in *The Tramp*. When he took the physical to become a Ranger, the doctor nixed his candidacy. "Sorry, son, you don't qualify. Next man," the physician barked. Rudder agreed with the rendering. Petty, however, refused to accept the exclusionary medical verdict. He started arguing, demanding either a more thorough Mayo Clinic–like medical exam or a simple second chance. "You've got false teeth," the doctor said. "You can't be a Ranger with false teeth. There's no way you can get in the Rangers without your choppers, son!" Petty, whose teeth had been knocked out playing football, went to see Rudder. The very fact that even after fracturing both legs he still wanted to be a Ranger showed true heart. Petty insisted to Rudder that it was unfair to disqualify him because of his dental malady. He ended his plea with a dramatic "Hell, sir! I don't want to eat 'em. I want to fight 'em." A smiling Rudder signed a form and handed it to Petty. "Here, take this back down there to the medics. You're in." Petty was assigned to F Company and would win the Silver Star at Pointe du Hoc for killing over thirty Nazi soldiers.

Although many men passed the physical, the unbearable Tennessee humidity caused many of them to throw in the towel. Dehydration was rampant. The sultry heat enveloped the would-be Rangers with brutal, unrelenting consistency. Uniforms would be drenched with sweat after only five or ten minutes of drilling in combat gear. Exhaustion was the normal state of being. "Rest," it turned out, was a word for the civilian effete. Staten Islander Herman Stein, a member of F Company, years later told of how his buddies were "dropping like flies" due to heat exhaustion. To survive the regimen he broke the rule of one canteen of water per long march. "On one extended break we stopped by a swampy section, filled up our canteens, added a couple of iodine pills, and presto—instant revitalization," Stein recalled. "This was the civilian soldier's way of thinking—you take orders to a degree but when it interferes with your rationalization, and you start performing like a zombie, then it's time to do something about it. Many times under fire we made up our own minds what to do."

On August 1, just as the 2nd Ranger Battalion began to grow accustomed to Camp Forrest, they were transferred to subtropical Fort Pierce, Florida. Located just north of Palm Beach on the Atlantic Ocean, it was the home of the U.S. Navy Scouts and Raiders School.* The muggy Tennessee woods, after all, were not the place to plan an amphibious Dieppe-like raid. True, you could build up endurance by jogging down rutted pine-forest roads in extreme heat or by scaling sheer mountainsides in a thunderstorm, but Camp Forrest lacked sufficient water for amphibious training. These Rangers were being groomed to eventually cross the English Channel in a daring invasion plan, known as Operation Overlord. They needed to acquire sea legs. They needed to learn how to navigate rubber boats, patch them up and inflate them while under fire. They needed to learn how to avoid the hideous pitfalls of seasickness, undertows, and rising tides.

Shortly after arriving in Fort Pierce by train, the Rangers made the acquaintance of so-called Navy frogmen, the underwater demolition experts who would be renamed SEALs

*As the 2nd Ranger Battalion headed to Florida, the 5th Ranger Battalion was activated at Camp Forrest.

nearly twenty years later, when John F. Kennedy was president. These aquatic frogmen had found this palm-tree-lined town located along the Indian River perfect for dives. One part of Fort Pierce, in fact, was already known as "Dynamite Point" because of its frequent use by the Navy Underwater Demolition Teams. The Rangers—after learning amphibious warfare tactics from the frogmen—were immediately put into an intense eleven-day class. Among their teachers were scarred 1st Ranger Battalion survivors of the bold invasion of Algeria and Tunisia. All of their training was leading to a final, all-out graduation drill: the mock attack, and capture, of humble Fort Pierce. "The Rangers held to a round-the-clock schedule," Ronald L. Lane wrote in *Rudder's Rangers.* "It seemed to boost morale. They practiced landing on rock jetties, loading and unloading into Landing Craft, Assaults (LCAs) from landing nets, and studying silhouettes on shore from positions at sea. They made company and battalion cross-country marches which included portaging of their rubber boats, making beachheads, infiltrating through beach-sentries, and attacking small installations. They learned how to send small beach-marking parties ashore from boats at sea."

Anybody who underestimates mock battles—thinking they're something akin to recreational paintball matches held by weekend warriors all across America—is dead wrong. At Fort Pierce chronic seasickness, sand-fly bites, sprained ankles, and heat exhaustion were all part of the fare. Most discouraging of all is what happened when they tried to cool off in the ocean water: schools of jellyfish attacked them. "Moans could be heard from several tents where a

Ranger's flesh had been stung by the touch of these creatures," Lane recalled. "Even calamine lotion did little to help relieve the sting." The Rangers interviewed in recent years, however, seldom mention the deprivations of Fort Pierce. What they're most proud of is that they successfully "captured" the town with only a few broken ribs as misfortunes. The end result of the 2nd Battalion's enduring Florida—and capturing Fort Pierce—was that they had passed their amphibious training. They were truly cohering as a unit.

On September 17 the 2nd Ranger Battalion were once again relocated, this time to Fort Dix, New Jersey. Since most of the men were from the East Coast, this would provide them with an opportunity to reconnect with their families—or so they thought. In truth, the rigors of Fort Dix were daunting. This was their last stop before being shipped off to Great Britain, and Rudder wanted to make sure his Rangers were in tip-top shape. Everything from calisthenics to sprint marches and range firing intensified. At night they rubbed black grease on their faces and practiced covert moonlit actions, presumably unnoticed by the locals. One survival drill had the men dropped off in an isolated New Jersey pine barren for a few days. They had to live off the land, shooting rabbits or stealing chickens to stay alive. A particular focus was given to scaling fifty- to sixty-foot-tall towers. For a few days the men were sent to Maryland for intelligence briefings. Together they had already made the history books: they established a U.S. Army record fifteen miles in a two-hour speed march. And then, they had their last hurdle: a complete physical.

Decades after D-Day, the surviving members of Rudder's Rangers had fond memories of Fort Dix. It was during this period that the enlisted men were promoted from private to private first class. Because they worked so hard during the day, Rudder gave his men night passes. They frequented such bars as the Shamrock Inn and Herby's Derby. Over pitchers of cheap beer they sang Perry Como, Bing Crosby, and George M. Cohan songs. Those more adventurous hopped the Greyhound bus for the seventy-four-mile trip to New York's Times Square, Greenwich Village, and Rockefeller Center.

The 2nd Ranger Battalion's last weeks stateside involved an accelerated course on what constituted the Third Reich. Most of the men were forced to learn basic German; all of them were instructed on how to use captured Nazi weaponry. In intelligence briefings they learned about the Gestapo, the French Resistance, and *Mein Kampf*. It all fell under the essential wisdom "know thy enemy."

Just before Thanksgiving, the men of the 2nd Ranger Battalion boarded the *Queen Elizabeth*, headed for the port city of Greenock, Scotland. The stately passenger liner, the largest steamship in the world, was the ideal troop transport because of its unsurpassed capacity and its unmatched speed. Before leaving the soldiers were all issued an official blue and gold Ranger patch—they had passed the six-week endurance course. To them it meant more than a diploma, license, or certificate. It represented their honor and it was worth dying for. Regular soldiers, however, called the patch a "Blue Sunoco" because it closely resembled the logo of the

Sunoco Oil Company. Jealous barbs aside, these Rangers were now more than a "band of brothers" of nearly five hundred strong; they were an elite fraternity of lethal killers.

While crossing the Atlantic, the Rangers served as military police, interacting with the few civilian passengers who for one reason or another were headed to Europe. The true purpose of their voyage was never revealed. Well-rested and battle-ready, they arrived in Scotland on December 2, 1943, and were quickly transported south. "Five hundred Rangers were greeted with 500 pairs of boxing gloves and a few miles of rope for our training on the cliffs of Cornwall," Owen Brown recalled. "We climbed the cliffs along the Cornish coast and hiked the hills inland, just for conditioning."

The home base of Rudder's Rangers was the town of Bude, England, located on the Atlantic coast of Cornwall. As champagne was uncorked on New Year's Eve, speculation was rampant among the Rangers. Another group of specially trained U.S. Rangers, the 5th Battalion, commanded by Lieutenant Colonel Max F. Schneider, would soon also arrive in Great Britain to train alongside the 2nd. Ever since mid-1942, Allied planning for an amphibious invasion of France had been in the works. But when? How many men would be needed? None of the Rangers knew. While the rumor mill ginned up various scenarios, the launch date remained top secret. The standard answer officers gave to the question "When?," however, was some variation of "Sooner rather than later." What the Rangers *did* know was that they would in all likelihood have to make a cross-Channel invasion sometime in 1944 or 1945. Meanwhile, Rudder's Rangers spent January and February 1944 scaling Cornish cliffs, firing at movable targets with automatic weapons, and kindly ingratiating themselves with the citizens of Bude. Instead of living in barracks, in fact, the Rangers were assigned housing with locals. (Some of the Rangers forged lifetime friendships with their host families.) "There were a pair of buddies in living quarters in private homes, and we paid the people we lived with so much a week, and we would get per diem every month to pay our expenses," 2nd Battalion Ranger Salva Maimone recalled. "When we traveled to a place to meet, we'd go in twos. We had a mess hall down in Bude, where we had our headquarters and we all gathered, and had our lunch, breakfast, and dinner there; then we'd go back to our private homes and do what we needed to do while we weren't training."

The oddest part of the Rangers' training, a reoccurring drill that often unnerved them, was when they were suddenly ordered to be in some far-flung British town at a designated hour. Promptness was mandatory. How they were to get to a Scottish hamlet or London suburb—each at least a day's drive away—was up to them. It would take initiative. That was the quality Rudder most admired. "*In-nit-a-tive.*" They could sneak on a train or bum a car ride or borrow a bicycle—it didn't matter. Getting to the designated destination at the desired time did. The objective, in retrospect, was obvious. These men, Rudder's Rangers, would soon have to learn to get around Occupied France with a compass once they climbed the cliffs of Normandy and were in German-occupied hostile territory. Initiative—what the dictionary defines as the determination or ability to begin or to follow through energetically with a plan or task or enterprise—was the quality that would make the difference between life or death, freedom or fascism.

The hazy time frame regarding the dangerous assault missions these Army Rangers trained for was cleared up in May 1944. A calm, methodical Rudder called his anxious men together in Cornwall and laid out the specific goal of the upcoming D-Day invasion. "We were called in not on an individual but on a very small unit basis and briefed," Ranger Jack Keating recalled. "We were taken into this room where they had a map or an overlay—it was like a map that had the beaches outlined, and we were shown exactly where we were going to land, what our job was going to be, and for the first (probably) four or five days after D-Day, if you survived." Likewise, Ranger Owen Brown claimed he would never forget the fateful day when the purpose of their grueling training became crystal clear. "We had miniature plaster layouts of the French coast, giving the location of beaches, cliffs, and the German defenses," he recalled. "They showed us pillboxes, barbed wire, minefields, underwater obstacles. They even told us the composition of the soldiers. There was supposed to be a cadre of Germans with Polish, Russian, and Hungarian men from captured countries working and fighting for the Germans."

Rudder, now a lieutenant colonel, was thrilled that the 2nd Ranger Battalion—comprising six companies with sixty-three men each—was chosen to play such a crucial role in Operation Overlord.* The rumors were over—reality was sinking in. His Rangers would be responsible for climbing the sheer cliffs at Pointe du Hoc and destroying the menacing Nazi concrete bunkers located four miles west of Omaha Beach on the coast of Normandy. The U.S. Army's intelligence branch knew that the Germans had a battery of six 155-mm guns perched on top of the Pointe, between Utah and Omaha beaches, in position to deliver devastating fire on both beaches as well as on naval targets. Their importance was not lost on the Allies, who had Pointe du Hoc on every target schedule on D-Day. With an estimated firing range of 25,000 yards (about 14 miles), taking out these guns could make the difference as to whether D-Day was a stunning success or a dismal failure. From their jutting crow's nest the Germans could inflict great damage to the Allies' armada of over 5,000 ships. Hitler had located his 716th Infantry Regiment and 726th Infantry Regiment to safeguard his precious guns; even though there were only about 125 of them, the Germans wouldn't give them up without a hellacious fight. Stationed behind Pointe du Hoc, however, were elements of the German 352nd Infantry Division, based near the town of Saint-Lô; they would in all probability quickly rush to the Pointe to thwart the Allied invasion once it commenced.

To neutralize these enemy guns, this elite group of about 225 2nd Ranger Battalion volunteers—Rudder's

*Companies D, E, and F totaled 189 Rangers. But Headquarters Company provided additional men for the assault on the cliffs, including medics and communications specialists. These units constituted the Boys of Pointe du Hoc, said to number 225 strong.

Rangers—had to surprise the enemy by scaling the 100-foot-tall promontory, even while under a barrage of Nazi fire coming from a concrete observation post. It was a daunting assignment. Erwin Rommel, now in command of Atlantic Wall troops, took limited precautions to secure the area but never believed Eisenhower would strike the Pointe. The other 275 Rangers—Companies A, B, and C—would attack elsewhere along the Normandy coast.

To make the Rangers' job easier, the U.S. Army 8th and 9th Air Forces would bomb the dickens out of the Pointe du Hoc area before the invasion began. Pointe du Hoc was so important that in the months preceding the invasion, aircraft returning from bombing runs in France and Germany were instructed to drop any remaining bombs from their missions on the Pointe rather than in the English Channel, as was customary. The natural fortress was heavily bombed throughout May during daytime hours, and on June 2, 3, and 4 during both day and night. With some 3,500 aircraft in their arsenal—including P-47s, P-51s, and B-25s—the 9th Air Force was the largest tactical air unit ever assembled. On D-Day proper, the 8th Air Force's B-17s and B-24s were assigned to abuse the Pointe du Hoc outcrop; their bomb tonnage created avalanches down the cliff and massive craters on top. At the very least the thunderous bombardment would send the Nazis scurrying for cover in their underground bunkers. This was welcome news. Help was also being provided by the RAF's 2nd Tactical Air Force. But tactical aviation support wasn't a life insurance policy. The Army Air Forces and RAF bombers couldn't just turn the Pointe into a leviathan bull's-eye. Nothing would undermine the invasion plan more than if the Nazis started suspecting—due to bomb-drop ratio—the location the Allies would try their invasion. So select German targets were bombed up and down the French coastline from Brittany to Belgium. These Atlantic Wall raids increased in frequency as D-Day neared. "The planning staff understood that, while air and naval support could cause great damage," historian Robert W. Black wrote, "only men on the ground could insure control."

Over the years World War II chroniclers like John Keegan, Stephen E. Ambrose, and Martin Gilbert have commented on how difficult an assignment these 225 Rangers were asked to perform at Pointe du Hoc. They were all combat rookies. They were being asked to go on what was essentially a suicide mission. When General Omar Bradley, in fact, asked Rudder to take on the assignment, the no-nonsense Texan gulped as he said, "Yes, sir." But as they looked at each other their eyes locked. They both knew the body count would be extraordinarily high. "No soldier in my command has ever been wished a more difficult task than that which befell the thirty-four-year-old Commander of this Provisional Ranger Force," Bradley wrote in his memoir, *A Soldier's Story*. "Lieutenant Colonel James E. Rudder . . . was to take a force of 200 men [*sic*], land on a shingled shelf under the face of a 100-foot cliff, scale the cliff, and destroy an enemy battery of coastal guns." The first time Bradley mentioned the mission to Rudder, the Texas colonel deemed it a cruel practical joke. Rudder later told Bradley, "I thought you were trying to scare me."

Sometimes human courage is a matter of old-fashioned

luck, good fortune, or impulsive response. It's not always an inherent, self-evident quality. But in truth, these Rangers lived by the maxim made famous by Branch Rickey, general manager of the Brooklyn Dodgers: "Luck is the residue of design." Praying wasn't enough to decimate the Nazis. You had to outthink and outhustle and outtrain them every day. While it's true that the most popular refrain of World War II troops would become the self-deprecating "I was just doing my job," Rangers believed differently. They trained to outperform expectations, particularly during the ferocious helter-skelter of battle. They were unflappable under pressure.

It's astonishing that as Jack Keating and Owen Brown, for example, studied maps and models of the Normandy coast and were told that they might be ordered to climb towering cliffs and attempt to destroy a battery of German guns emplaced in reinforced concrete casemates, they didn't go AWOL. Instead of with nausea, they greeted the news enthusiastically, eager as they were to liberate Europe. It's not hyperbole, in fact, to claim that their collective courage—and love of country—was of the Homeric kind. The odds of survival had to be, at best, fifty-fifty. Look at it this way. Climbing up the side of El Capitan in Yosemite National Park is a mountaineering feat in its own right. But doing it while the German Army is raining machine-gun fire and mortar rounds at you would be either insane or Herculean; take your pick. That all of these Rangers were willing to volunteer for such a deadly mission—no questions asked—is awe-inspiring. "As we prepared for the invasion, we were all aware of the dangers, but at least as far as I was concerned, I didn't really think about the danger in a personal way," Ranger Gerald Heaney recalled. "It was as if we were so well trained and so well prepared that nothing could stand in our way."

With accelerated resoluteness Rudder began further training his Rangers for the contingencies of the D-Day invasion. On May 6 a Provisional Ranger Group was officially formed; it was ordered to attach itself to the 116th Infantry Regiment, 29th Infantry Division. In charge of this Ranger group, including both the 2nd and 5th Battalions, was Colonel Rudder. According to Lieutenant George Kerchner, Company D, Rudder ordered them to climb the "most difficult cliffs that they could find." Rudder located the ideal ones, the closest in topographical proximity to Pointe du Hoc, at Swanage on the coast of Dorset. (Other steep terrain was climbed on the Isle of Wight.) They were 200 to 250 feet high. The eager Rangers boarded British LCAs and then stormed the docile cliffs. Veteran British commandos, with combat experience under their belts, helped with the accelerated amphibious training. At Swanage the Rangers also experimented with a vast array of equipment for escalade, including steel extension ladders. Due to their inherent lightness, rope ladders, however, were the preferred climbing tool. DUKWs* were also added to the invasion mix. They would land on the shale beach at Pointe du Hoc, raise a ladder high into the air, and allow Rangers to climb to the top rung, unloading machine-gun fire on the unsuspecting Nazi lookouts.

*According to *Brassey's D-Day Encyclopedia,* the name DUKW was an acronym based on manufacturing designations: D for model year 1942, U for amphibious, K for all-wheel drive, and W for dual rear axles. DUKWs were informally called Duck boats.

LCA-722 participates in pre-D-Day training exercises in Great Britain. (U.S. Army Military History Institute)

It was at Swanage that the 2nd Rangers learned what they were allowed (ordered) to bring with them on their seaborne assault mission. Because they were going to have to survive for days or even weeks on their own if they reached the top of the cliff, the Rangers were expected—by a March 1, 1944, order—to have the following provisions:

a. Individuals in the assault will carry the minimum equipment for the assigned mission. The combat pack will contain only the following: 2 rations (1 "K" and 1 "D"); spoon; 1 suit underwear; 2 pair socks; 1 raincoat (if not worn); 1 towel; and 1 toilet kit.

b. Individual rolls containing blankets, shelter halves, and mess kits will be provided as soon as the tactical situation permits and in accordance with administrative order.

c. All units will carry organic loads of POL [petroleum, oil, lubricants], water, and ammunition insofar as vehicles will permit. Three (3) days' ration (type "k" or "c") will be taken for emergency, not to be counted in total.

Left to their own wits, the Rangers knew they would succeed. Their confidence was that high—or so they feigned. What the Rangers were fearful of, however, as they physically *and* psychologically prepared to take Pointe du Hoc, was the hungry artillery the Germans had in place along the Atlantic Wall. The guns on Pointe du Hoc, ironically, weren't German-made, a fact often overlooked by D-Day scholars. The Nazis had captured these 155-mm guns from the French Maginot Line in 1940. (The French-designed Grande Puissance Filloux were manufactured to fire four rounds per minute with a muzzle velocity of 735 meters per second.) History documented that previous English-bound cross-Channel invasions—the Norman Conquest, for example—could be deemed successful. No soldiers, however, had ever been asked to directly charge enemy guns of such brutal magnitude. Such weapons simply didn't exist in the sixteenth century, for instance. But make no mistake about it: all of Rudder's Rangers understood their daring high-wire act was something new in the annals of warfare and was absolutely essential if Operation Overlord was to become a military reality and not just a pipe dream of Roosevelt and Churchill. According to Rottman, Rudder's Rangers understood that the German guns' "ability to deliver flanking fire on both Omaha and Utah Beaches" was endangering "the D-Day landings."

As D-Day approached, the 2nd Ranger Battalion entered

its final seven-day course at the Army Assault Training Center in Braunton, Devonshire, a combat simulation facility overlooking Barnstaple Bay where the Fabius-7 full-scale pre-invasion exercises were held. The drills closely approximated what the Rangers could expect at Pointe du Hoc and Pointe et Raz de la Percée, where Company C would land. (There were no German big guns at the latter promontory, located three miles east of Pointe du Hoc, but the Nazi small-arms fire was expected to be intense there.) For weapons, the Rangers were given everything from bazookas to bangalore torpedoes and thermite grenades. Single-channel radios were provided; after all, once they landed on the beach and the Germans were shooting at them, communication would be *the* essential ingredient for ultimate success. And, the Rangers were reminded, they would not be alone. Operation Overlord called for over 6,000 naval vessels, including 138 warships to provide cover for the first-wave outfits like the 2nd Rangers. Advance intelligence had also reported that the D-Day beaches of Omaha, Utah, Sword, Gold, and Juno were booby-trapped with land mines and barbed wire. At Braunton they were taught how to carefully avoid and deactivate such life-threatening obstacles.

It was during these springtime drills that the 2nd Rangers learned definitively about their D-Day landing vessels. In the United States, they were designated Landing Craft, Vehicle, Personnel (LCVPs) or Higgins boats (after their New Orleans–based shipbuilder, Andrew Jackson Higgins). During World War I, amphibious forces used river barges or old whaling boats for landing craft. "If Higgins had not designed and built those LCVPs, we never could have landed over an open beach," Eisenhower believed. "The whole strategy of the war would have been different." It wasn't until the late 1930s that Higgins Industries designed LCVPs, which could land men *and* equipment on a beach. Once the boat was ashore, the troops on board would leap over the side or sprint out via the equipment vehicle ramp. Called by Hitler America's "new Noah," Higgins had designed these vessels especially for amphibious assaults. The boat was crafted mostly of plywood; the only metal part was the front ramp door, which served as a barrier against enemy fire and then dropped open. Holding up to thirty-six men, these tough and adaptable boats could operate in less than two feet of water, and their curved fronts and guarded rear propellers allowed them to pull onto land and still be able to maneuver back out to sea. By September 1943, 12,964 of the U.S. Navy's 14,072 vessels had been designed by Higgins in the warehouse behind his St. Charles Avenue showroom—meaning that 92 percent of the Navy was a Higgins navy.

The 2nd Ranger Battalion, however, would use a slight variation of the Higgins boats for their D-Day landing. Because the Rangers had been in Great Britain for nearly six months and had trained with commandos, they had grown accustomed to practicing assault maneuvers in forty-one-foot-long Landing Craft, Assault boats, the British equivalent of a Higgins boat. Both models featured lowering ramps and had two engines. The only real design differences between an LCA and a Higgins boat were gunwales and light armor on the sides of the former. "That made the LCA slower and heavier," Stephen E. Ambrose wrote in *D-Day, June 6, 1944: The Climactic Battle of World War II.* "Which meant that the LCA rode lower in the water than the LCVP." It would take ten LCAs, fully equipped with three pairs of rocket mounts, to carry Companies D, E, and F to the narrow beach in front of Pointe du Hoc. Like the LCVPs, the LCAs could carry up to thirty-five men. They were, however, too flimsy to effectively transport vehicles. As vessels go, the LCAs were quite silent, designed to beach without revealing their positions because of noise. More than 480 LCAs were deployed on D-Day.

The great innovation, the one that made the liberation of Pointe du Hoc possible, was not LCAs. It was a new mountaineering device. Fastened on the front of select landing craft headed toward Pointe du Hoc would be a specially designed rocket containing a grappling hook with ropes connected to it. When the LCAs hit shore, these rockets would be discharged straight into the cliff. According to the strategic plan, the fired grappling hooks—an iron shaft with claws at one end used for grasping or holding—would penetrate into the Pointe; this would allow the U.S. Rangers to climb up the rope ladders that dangled down to the beach.

Out of all the first-person accounts about the Boys of Pointe du Hoc, the most detailed was an article written shortly after D-Day by Lieutenant G. K. Hodenfield that appeared in the *Saturday Evening Post* on August 19, 1944. A writer for *Stars and Stripes,* Hodenfield understood that Pointe du Hoc was "target No. 1," and wanted to cover the action. He had joined the 2nd Ranger Battalion only three days before Operation Overlord commenced. He quickly learned that the rope ladders with grappling hooks which were to be fired over the cliffs by rockets were the "secret weapon" of the assault. "The grapnels were to bite into the bomb-blasted earth of Pointe du Hoc, and when the slack was taken up, the ladders would be ready to climb," he explained to readers. "The entire success of this operation depended on those ladders. Pointe du Hoc is accessible from the sea only by scaling the cliffs, and the Germans, believing that not even 'military idiots' would dare to come from that direction, had placed all their defenses facing inland."

The Pointe du Hoc headland is a remarkable natural and strategic spot. It jabs into the English Channel like a cursed dagger. But to be successful, the attack of the U.S. Army Rangers' 2nd Battalion on the Pointe's overhanging cliffs had to rely upon the essential element of complete surprise. As Hodenfield noted, it was reasonable to presume that the German defenders would not expect a Ranger battalion to attempt to scale the cliffs. After all, not even Rommel could have envisioned grappling hooks fired from special mortars installed on a fleet of newfangled landing craft. It was precisely because the Nazis would not anticipate such a bizarre invasion strategy that Supreme Allied Commander Eisenhower planned it: a plausible but high-risk military gambit with everything at stake. "The Allies were invading a continent where the enemy had immense capabilities for reinforcement and counterattack, not a small island cut off by sea power from sources of supply," U.S. naval historian Samuel Eliot Morison wrote in *The Invasion of France and Germany: 1944–1945*. "Even a complete pulverizing of the Atlantic Wall at Omaha would have availed nothing if the German command had been given twenty-four hours' notice to move up reserves for counterattack. We had to accept the risk of heavy casualties on the beaches to prevent far heavier ones on the plateau and among the hedgerows."

As D-Day approached, Colonel Rudder—like all of his Rangers—turned introspective. With death an imminent possibility, afterdinner hours were consumed by writing heartfelt correspondence to loved ones. For security reasons, no mention of D-Day was allowed in these missives. Reading dozens of them decades later, one realizes that most of the Rangers wanted to remind family and friends that they were all right, that life was good and precious and, in their cases, possibly short. On June 1, for example, just five days before D-Day, Rudder wrote his wife, Margaret, and his children an upbeat, sweet note meant to show both patriarchal concern and inner strength. Shortly after composing it, he ordered his Rangers to board their transport vessels, which would wait in the devilish English Channel until they were ordered to strike France:

> ***My darling wife and kids,***
>
> *About now you are depositing Bud and Liz with Jewel and you are greatly excited over your trip to S.A. [San Antonio] tomorrow. Bud, you and Liz are getting set to have lots of fun playing with Billy and Lucille, while mommie shows the [damn yankee] the sights of Texas. I am equal pleased with my immediate future and that is kicking the pants off those so and so's that has [*sic*] separated me from the three persons that are most dear to me. We are really ready for the job—no one has ever had the privilege of leading a fine[r] bunch of men and officers into battle than*

I will have. Tell Alice she can well be proud of the Captain. He did a splendid job for me as adjutant and is doing equally as well in his present job. I am depending on him a great deal in the task ahead of us.

*Bud, ol Pal, I have been looking at your pictures that Smitty made of you, Liz and mommie and I must say Liz rather beat you in posing for pictures for her daddy. You surely aren't afraid to have your picture made are you? Ol pal you must always be big and brave and do the things just as mother wants you. Liz, you certainly were sweet in your little blue dress. Is it the one bought for you. Bud you were quite a sport in your stripped [*sic*] coat. Honey, the thinness of your face and neck gives me much concern—I do hope [you] are well and are putting on some weight. Remember our goal has only been half way reached. I want you to be in fit condition to get on with the job next summer when I return. Hold on to your help and don't let the snobs of Brady get her away from you. My darling you are more high powered than either of the ones you mentioned so stay with them.*

Give Alice my love and you two have enough fun for the Capt. and myself. Good night ol pal, Daddy's baby and my darling

ALL MY LOVE, EARL

Colonel Rudder wasn't the only officer taking pen to paper in the countdown hours before D-Day. General Eisenhower, besides writing his wife, Mamie, sent this message to all Americans participating in the D-Day invasion of Normandy:

Soldiers, Sailors and Airmen of the Allied Expeditionary Force!

You are about to embark upon the Great Crusade, toward which we have striven these many months. The eyes of the world are upon you. The hopes and prayers of liberty-loving people everywhere march with you. In company with our brave Allies and brothers-in-arms on other Fronts, you will bring about the destruction of the German war machine, the elimination of Nazi tyranny over the oppressed peoples of Europe, and security for ourselves in a free world.

Your task will not be an easy one. Your enemy is well trained, well equipped and battle-hardened. He will fight savagely.

But this is the year 1944! Much has happened since the Nazi triumphs of 1940–41. The United Nations have inflicted upon the Germans great defeats, in open battle, man-to-man. Our air offensive has seriously reduced their strength in the air and their capacity to wage war on the ground. Our Home Fronts have given us an overwhelming superiority in weapons and munitions of war, and placed at our disposal great reserves of trained fighting men. The tide has turned! The free men of the world are marching together to Victory!

I have full confidence in your courage, devotion

to duty and skill in battle. We will accept nothing less than full Victory!

Good Luck! And let us all beseech the blessing of Almighty God upon this great and noble undertaking.

DWIGHT EISENHOWER

Besides having commanding officers pass out this morale-boosting message, Eisenhower wanted it airdropped and handed out all over Europe. Just about every U.S. soldier in the European Theater eventually read this one-page flier. Another urgent message Eisenhower wrote, however, which *none* of the D-Day soldiers got to read, was a top secret June 5 statement in which he took full personal responsibility if the invasion was a disaster. During the days before D-Day, Eisenhower kept it in his wallet. "Our landings in the Cherbourg-Havre area have failed to gain a satisfactory foothold and I have withdrawn the troops," he wrote. "My decision to attack at this time and place was based upon the best information available. The troops, the air, and the Navy did all that bravery and devotion to duty could do. If any blame or fault attaches to the attempt it is mine alone."

Although Colonel Rudder had never issued such a defeatist message, he was, like General Eisenhower, ready to take full blame for whatever went wrong at Pointe du Hoc. Like any good commanding officer, Rudder considered the 225 Rangers he helped train in Tennessee, Florida, New Jersey, and Great Britain to be his extended family. Together they would fight—win, lose, or draw—united in unflinching purpose.

Command post at Pointe du Hoc with U.S. flag showing, June 8, 1944. (Courtesy of the Eisenhower Center for American Studies, University of New Orleans)

3

CLIMBING THE CLIFFS, DESTROYING THE GUNS

Once Colonel James E. Rudder mailed his letter home on June 1, he ordered Companies D, E, and F—then at Marshaling Area D-5 in Dorchester, England—to board their British transports, HMS *Ben Machree* and *Amsterdam.* Meanwhile, Companies A, B, and C were ordered onto the *Prince Charles.* Their main port of debarkation was Weymouth. As the Rangers boarded their transports their primary concern was eluding the German Luftwaffe, whose planes were constantly buzzing over the Dorchester area taking reconnaissance photographs—and sometimes dropping bombs. The men in Companies D, E, and F were now part of a ten-LCA-strong team called Task Force A who were to capture Pointe du Hoc at H-Hour 6:30 A.M. on D-Day, then scheduled for June 5. But, of course, poor weather caused a postponement until June 6. The stakes were high. All of the 2nd Rangers looked at one another as if death were at their doorstep. But that was only for a flash second or two. Somehow, due to the combination of training and prayer, they collectively conjured up the courage to match Eisenhower's words, to tell themselves, "Okay, let's go." Indecision was an impossibility.

Many of the Rangers who boarded the transports were sick; the culprit was food poisoning from a potful of contaminated hot dogs. "It was bad enough having to contend with one's nerves before battle, now the food which they had been served was rancid," JoAnna McDonald recounted in *The Liberation of Pointe du Hoc.* "The whole battalion was laid low, and many began to doubt if they could go ahead with their mission. Some even suspected that their food had been tampered with."

Sitting aboard their transport ships, waiting for General Dwight D. Eisenhower's golden word "go," the Rangers whittled away time. They played low-stakes poker, sang Tin Pan Alley songs, and sipped Irish whiskey. A ukulele was passed around, although nobody strummed it well. A few preferred to quietly play bridge. Cigarettes were chain-smoked. Sleep was hard to come by while in limbo.

Mandatory intelligence briefings offered little new information about the status of the German guns, but they were a diversion from *the wait.* Pocket maps of the Norman countryside were studied carefully, as if the Rangers were young surveyors for *National Geographic.* Weighed down with equipment, they tried on gas masks and made sure they had an orange diamond Ranger decal on the backs of their steel helmets. On their left upper sleeves were their hard-earned Ranger patches. While all the men were comfortable in their green infantry pants, their jackets were a different story entirely. As a precaution against German blister gases, the jackets had been laced with a protective chemical, one that made them feel awkward to move around in. They were dressed in fatigue uniforms, carrying two grenades each and with an M1 rifle slung over their shoulders. Each of the three Ranger companies in Task Force A carried with it a couple of light mortars and special thermite grenades to be used once on top of Pointe du Hoc; blowing up German pillboxes and big guns was, after all, their main mission.

A civilian soaking up the pre-invasion atmosphere inside the *Ben Machree, Amsterdam,* and *Prince Charles* would have thought he or she had entered floating ammo depots. Everywhere one looked were Tommy guns, Browning automatic rifles, and buckets full of grenades. Young men who should have been dating girls or playing college football paced around the tight vessel quarters with rounds of ammunition crossing both shoulders and forming an ominous X across their chests. Because all the officers and most of the enlisted men were wearing paratrooper boots, the floor constantly rumbled. Sergeant Frank South, Headquarters Company, spent his pre-invasion hours packing plasma, suture material, sulfa-based antibiotics, and other medical supplies that would be essential once the heavy fighting commenced. A

Nebraska native, he learned how to be a medic when assigned to the 106th Division. Father Joe Lacy, a rotund, happy-go-lucky Catholic priest, held prayer sessions as the fateful invasion hour neared. "One of the things that I remember him saying to all of us on D-Day, at least to the people aboard the ship, that always stuck in my mind," Lieutenant George Kerchner recalled, was when "he said, 'When you land on the beach and you get in there, I don't want to see anybody kneeling down and praying. If I do I'm gonna come up and boot you in the tail. You leave the praying to me and you do the fighting!' "

Within a few days, however, the long wait was over. Shortly before 4:00 A.M. on June 6 the intercom on the three transport ships suddenly blared, "Rangers! Man your craft!" This was it, D-Day was upon them. Bad weather be damned—General Eisenhower, at 0330 hours on June 5, had decided the weather was now in their favor. He greenlighted Operation Overlord (June 6) by shouting "Okay, let's go." With those three simple words, the largest seaborne attack in history—over 100,000 U.S., British, and Canadian soldiers strong—was about to commence. All 225 of the 2nd Ranger Battalion's Companies D, E, and F washed off sleep and checked their equipment. Those not too seasick or food-poisoned gulped down pancakes, guzzled citrus drinks, and quietly walked onto the waiting LCAs, manned by British sailors, which would be lowered into the Channel once they were full of combat-ready Rangers. "All aboard the Hoboken Ferry!" one of the men famously shouted. This brought about a round of hearty, spontaneous laughter, but nervousness was unmistakably in the air. The approximately

275 men of Companies A, B, and C were likewise jarred awake on the *Prince Charles;* they too breakfasted and then boarded the small fleet of LCAs. Each LCA carried about thirty-five men.

A purplish, misty dawn was upon the Rangers when, after the transports sailed most of the way across the Channel, their LCAs were dropped into the turbulent waters. They headed straight for the Norman coast as soft light was emerging out of the darkness. The dark blue-gray water was cold. Waves ranged from two to four feet. If you fell overboard, hypothermia was almost guaranteed. A cutting breeze had put a real sting in the air. Patches of low-lying clouds mildly obstructed vision. Everything was damp, dismal, and dreary. A minute in the English Channel that morning on an LCA seemed like an hour. You could almost hear the seconds tick. The objective for the day for Companies D, E, and F was straightforward enough: destroy the six 155-mm German guns with a fourteen-mile range emplaced in reinforced concrete casemates at Pointe du Hoc. But in warfare objectives can change every two or three minutes. The important thing was to be hyperalert. Daydreaming was a sure avenue to death.

The Rangers of Company C—three officers and sixty-five men—were the first to disembark from the *Prince Charles,* in LCAs headed straight into Pointe et Raz de la Percée. British sailors helped these Rangers into the landing craft, which were hanging and swaying in the davits. They were ten miles away from the French coast. Unlike the Higgins boats, which were first lowered into the water and then loaded, the British LCAs were crammed with the soldiers

A French 155-mm gun. (U.S. Army Military History Institute)

before the boats were lowered into the choppy water. "Down went the landing craft in the davits," remembered Lieutenant Sidney Salomon, one of the platoon commanders. "All conversation had come to a halt as if everyone had suddenly become mute. All were tense. Now a loud smack as the bottom of the landing craft hit the water. The first landing craft away from the transport began to circle, waiting for the second craft to maneuver into position. When both were in their assigned positions, the parallel trip in the direction of the shore began. We were two small bobbing objects in the choppy waters of the Channel."

A terrible omen occurred only a few minutes after all the LCAs were dropped into the water. Assault craft LCA-860, due to either poor engineering or too much weight, overturned. Captain Harold "Duke" Slater and twenty-five of his men from Company D, all loaded down with heavy equipment, started sinking to the bottom of the English Channel. They tried to rip off boots and backpacks, anything that would enhance their ability to float to the surface. Four men drowned, and the others were rushed back to an English hospital after their rescue. This fiasco eliminated more than one-third of the sixty-five-man force of Company D. LCA-914, a supply craft, also sunk, killing most of the men. Suddenly, the 225 men of Pointe du Hoc numbered closer to 180. "There was nothing we could do to help those poor guys. Just say a little prayer that they would be picked up before they froze to death," a *Stars and Stripes* reporter riding with the Rangers wrote. "We all wanted to help but the success of our mission was too vital, and the Rangers knew they were expendable."

The men on other LCAs, experiencing serious leaks, had to start bailing their vessels by using their combat helmets as buckets. Their teeth chattering in the cold, the Rangers on these still seaworthy LCAs let out a collective gasp of despair when they discovered an even more serious problem: the all-important climbing ropes were soaked, and the weight of the water would mean that the rockets firing the grappling hooks wouldn't hit as high into the cliffs. The debacle of the infamous Dieppe raid flashed through some of the Rangers' minds as they huddled together in the LCAs like nervous cattle jammed inside a large phone booth. Years later, these moments aboard the LCAs conjured in the surviving Rangers the most tears. They had already watched

Ropes being fired toward the cliffs by an LCA in the English Channel on D-Day. (U.S. Army Military History Institute)

helplessly as friends drowned, engulfed by the hungry sea, and now they contemplated that the waterlogged ropes made reaching the cliff tops an even more difficult proposition.

No matter how many oral histories are collected about Rudder's Rangers, it's still impossible to know precisely what each man felt as they crossed the English Channel. There was not one set of war experiences from that day. Memory is episodic and details are easily forgotten. Every man crowded onto those LCAs, however, was pumped up, adrenaline coursing through his veins. Some men recalled daydreaming about family back in Ohio, New Jersey, or Pennsylvania. Others remember seasickness, raging sweats, or the counting of rosary beads. Jokes were a welcome relief. Compulsively recleaning weapons was commonplace. To break up the nervousness, and keep morale high, First

Sergeant Len Lomell, Company D, started placing hundred-dollar bets on which LCA crew would be the first up the cliffs. Most of the Rangers, like Lieutenant James Eikner of Mississippi, preferred decades later to simply recall their collective grace under fire. "When we went into battle after all this training there was no shaking of the knees or weeping or praying," Eikner matter-of-factly noted. "We knew what we were getting into. We knew every one of us had volunteered for extra hazardous duty. We went into battle confident. . . . We were intent on getting the job done. We were actually looking forward to accomplishing our mission."

Donald Scribner, from Company C, was on the *Prince Charles;* his unit's specific mission was to climb the cliffs of Pointe et Raz de la Percée, located three miles east of Pointe du Hoc. The C Company, isolated from the rest of the 2nd Battalion force, bravely fought the bewitching tidal current, praying they wouldn't sink. "I remember quite well going across the English Channel," he later recalled. "It was very rough. The waves were very high. We were about ten miles from shore when Colonel Rudder came down and talked to us prior to loading up the LCAs. He had this comment to make to us: 'Boys, you are going on the beach as the first Rangers in this combat in this battalion to set foot on French soil, but don't worry about being alone. When D, E, and F take care of Pointe du Hoc, we will come down and give you a hand with your objective. Good luck and may God be with you.' "

Decades later, Scribner reminisced about what it was like for Company C as they approached Pointe et Raz de la

Percée from the Channel like torpedoes heading directly into shore. "Suddenly there were splashes around the craft, and white, cascading water," he recalled. "Then concentric circles, as shells landed in the water in the vicinity of the landing craft. Sharp pings of bullets against the steel hull sounded, as defending Germans fired their automatic weapons directly at the landing craft." Instinctively, the Rangers crouched low in the LCAs and the coxswain propelled the throttle forward. The torrent of bullet pings on the hull, however, continued unabated. "I called out the words, 'Get ready,' which were passed along to those in the stern of the craft, and everyone inched forward just a little bit," Scribner recalled. "The moment the ramp had dropped down, automatic weapons and rifle fire sprayed the debarking Rangers, killing and wounding several men. The second man was hit by a bullet from a German entrenched at the top of the cliff. I reached over and pulled him clear of the craft, just before the heavy steel-hulled craft would have steamrolled him. The Rangers waded to the shoreline and started across the sand, striving to reach the cover of the base of the cliff. Machine-gun, small-arms fire, and mortars cut down some of the men as they attempted to run across the sand."

Quickly the men of Company C gathered at the base of the cliff and returned fire, hoping to pick off the German snipers ensconced above. Captain Ralph Goranson, the company commander, ordered the Rangers up the ropes. "Lieutenant [William P.] Moody, along with Sergeant Julius Belcher of Schwartz Creek, Virginia, and Otto Stevens of New Castle, Indiana, scaled the cliffs," Goranson recalled.

"These three men free-climbed about a ninety-foot cliff that was partially an incline and then straight up the last fifteen or twenty feet. We gave them covering fire from down below to keep the Germans off their back. The last ten or fifteen feet they chinned themselves up with their trench knives and secured a series of toggle ropes from the barbed-wire emplacement up there, so the rest of us could immediately move over into this position and climb up the cliff and get into the area around the fortified houses. Lieutenant Moody, immediately when he got topside, killed the officer in charge of the Germans in the fortified house. We found the rest of the area honeycombed with dugouts, trench systems. And immediately Lieutenant Moody and his men dispatched teams to go and clear up this area; they were followed immediately by the 2nd Platoon with Lieutenant Salomon, and it was right here that we lost Lieutenant Moody. He was downed by a sniper."

That left Lieutenant Salomon to lead the forces up the gnarly cliffs of Pointe et Raz de la Percée. Once atop he ordered Rangers in groups of twos and threes to capture the snakelike German trenches and dugout holes. With great caution the Rangers then surrounded the fortified house at the top of the cliff that was the nerve center of the Nazi base camp. "This was put out of action by Sergeant Belcher who threw in a white phosphorous grenade," remembered Captain Goranson, "and when the Germans came out, they were sent to heaven by Sergeant Belcher's gun."

Every Ranger interviewed over the years has vivid memories of German hand grenades and machine-gun fire. "Ultimately, I made it to the top, and spotted a series of trenches some twenty-five yards distant," Salomon went on in his oral history. "I pointed, indicating all should run to the trench. We all ran and leaped in the trench fully prepared to take sole possession, and opposition from the foe soon ended, and I looked around to see who was still with me. Nine remained from the thirty-nine that had been jammed in the landing craft."

While Company C attacked Pointe et Raz de la Percée, the small flotilla of nine LCAs with Companies D, E, and F navigated the heavy seas, headed directly to Pointe du Hoc beach. For these companies—the now celebrated Boys of Pointe du Hoc—things continued going terribly wrong. As the LCAs approached the coast, Colonel Rudder noticed that the promontory to which they were headed was *not* Pointe du Hoc, but Pointe et Raz de la Percée, where Company C was already engaging the Nazis. The swift current had sent them off course, about three miles east of Pointe du Hoc. Rudder quickly had the helmsman of LCA-888 turn right ninety degrees to head parallel to the coast, running a gauntlet of German small-arms fire. "Earl always believed if they had hit the beach on time, they could have taken Pointe du Hoc without firing a shot," Rudder's wife, Margaret, later noted. "But no matter how much went wrong, he and his men still succeeded."

The other LCAs followed Rudder's lead vessel in a straight line, approximately two hundred yards off the Norman shore. They were in a difficult—but not hopeless—situation. The important thing was not to panic. The good news was that the British destroyer *Talybont* and the American destroyer *Satterlee* were still firing away at the surprised Nazi lookouts. "I can remember when the first small arms hit our boat and it made a noise and somebody said what it was," Eikner recalled. "I looked and there was a little round hole through one of the rope boxes and I said, 'My God, these guys are playing for keeps,' and so we all got down. We had been standing up except for those who were bailing water, so we all ducked down. The Germans were taking us under fire like shooting ducks in a tub and it got worse as we got closer to the Pointe."

Suddenly seeing the promontory they had studied in reconnaissance photographs looming in front of them like Oz—a surreal, rocky outcropping from which German soldiers fired directly at their vessels—caused the most even-keeled Rangers to gulp. Contained anxiety spread. They weren't being asked to merely climb the cliff; it fell upon them to take the mountain, so to speak. Each man hoped, and many prayed, that U.S. Army Air Forces and RAF planes had been bombing the daylights out of the Atlantic Wall. Secondhand reports were that their hundreds of sorties had been highly successful. But who knew for sure? At least the billowing clouds of smoke rising above the coastline were testimony that *something* had been hit. Maybe the German guns had already been taken out? Maybe not?

The British coxswain's navigation blunder caused a thirty- to forty-five-minute delay in getting to Pointe du Hoc. By the time Rudder detected the error, precious time had been wasted and his men were under fire as they ran parallel to the shore. The narrow window to signal for the offshore force to follow had almost expired: time was of the essence. As nine LCAs—668, 722, 858, 861, 862, 883, 884, 887, and 888—approached the east side of Pointe du Hoc under a small-arms barrage, the Rangers of Companies D, E, and F were poised to attack. If the Rangers got control of the Pointe, two specially loaded backup LCAs would land on the beach with extra rations, mortars, and equipment. Then the other two Ranger companies—A and B, along with the entire 5th Battalion—would follow them ashore. These backup Rangers were offshore marking time, waiting to attack the Pointe. If the attack failed, or if the success signal did not come in time, they would go to nearby Omaha Beach and march up the Vierville draw to the coast highway, attacking the German fortifications at Pointe et Raz de la Percée and then move against Pointe du Hoc from the rear.

The secondary objective, after the coastal guns were decommissioned, was to cut the road running behind Pointe du Hoc from Ouistreham to Grandcamp. "If their comrades at Pointe du Hoc had succeeded, then the Rangers from Omaha Beach would link up," historian Jonathan Bastable explained in *Voices from D-Day;* "if they had failed, then the Omaha Rangers would attempt to capture the big guns from the landward side. So there were three separate chances to take the six great guns out of the battle: the bombardment from the air; the assault from the sea; and the smash-and-grab raid across the land." Bastable's explanation is wrong, in part. He suggests that the original plan was for a Ranger landing at Omaha Beach. That was to take place only if the attack at the Pointe failed, or had not succeeded in the time allowed. The original plan was for all the Rangers to assault up the cliffs.

As Rudder's Rangers approached Pointe du Hoc, firing their grappling ropes into the cliffs, the battleship *Texas* constantly roared rounds of 14-inch shells at the perched enemy forces. The explosions were deafening. The *Texas* was one of the most illustrious battleships of World War II. Before Pearl Harbor the vessel had received the first commercial radar in the U.S. Navy, followed by brand-new antiaircraft batteries and radio equipment. The venerable *Texas* had participated in World War I and was considered old in 1944, yet its new on-deck technology caused it to be designated the flagship of the U.S. Atlantic Fleet. Throughout World War II numerous German submarines tried to torpedo the *Texas* but were unsuccessful. And it was the *Texas*—providing gunfire support for the assault on Morocco—that dropped off the twenty-six-year-old war correspondent Walter Cronkite as he began his assignment for CBS.

Decades later, almost all the men of the 2nd Ranger Battalion speak of the withering fire of the *Texas* in near reverential tones. When they were storming the cliffs at Pointe du Hoc, the *Texas* was blasting away at the German defenses with its ten 14-inch guns, positioned in five turrets.* The battleship, USS *Satterlee,* and HMS *Talybont* were the best friends the U.S. Army Rangers' 2nd Battalion had.†

There was, however, one downside to this naval support—and it was a big one. As the Rangers landed on the

*After D-Day, the *Texas* was hit twice by German coastal defense artillery near Cherbourg and went in for repair. Eventually the *Texas* was sent to provide gunfire and antiaircraft services for the Iwo Jima and Okinawa landings.

†A second destroyer, USS *Harding,* relieved *Satterlee* late on D-Day and continued to provide direct support to the Rangers, but its action was not recognized until the war's end.

500-yard-long, 30-yard-wide beach in front of Pointe du Hoc, they were catching gunfire from both sides. In addition, some Rangers were killed or wounded from friendly fire. Surviving Rangers tell of dead on the beach, some missing limbs, made lifeless in seconds by beach obstacles and raining German bullets. Lieutenant George Kerchner, D Company, later recalled the "terrifying sound" the 14-inch guns of the *Texas* made. "Of course they were passing far over our heads, but we were close enough to hear and feel some of the muzzle blasts," Kerchner said. "I recall as we got perhaps halfway in, one of the rocket-firing craft that was off of Omaha Beach fired their . . . rockets. This was also a terrifying thing; I think there were a thousand or more rockets on these landing craft, and they fired in salvos of maybe ten or fifteen at a time. It was just one continuous sheet of fire going up from this rocket-firing craft. . . . I remember wondering how could anybody really live on the beaches with all of this fire that was landing there."

Company E's Salva Maimone remembers how his experience on his transport ship came flashing back into his mind as his LCA's ramp opened and he stormed the cliff, the rockets having already fired the ropes. "On the boat, we had a little conference with the officers telling us how dangerous this mission was and we were facing great odds. The way they put it to you . . . with the odds you had there, it was as if you were ready to go to an electric chair, because you didn't have any chance. And it had you feeling tight all the time you were on the boat. To try to forget about it, we played cards and tried to forget everything. But it went on, and [men] were laughing and carrying on, and the officers said everyone that even gets close to the cliff ought to get an award."

As the Rangers waded ashore at Pointe du Hoc, the logistical dilemmas caused by the delay were obvious. For one, the tide was quite high, and if the equipment-heavy Rangers didn't climb the cliff immediately, they would drown. "My buddy, Paul French, he would not put the strap on the back of his head or under his chin to hold his steel helmet on," Frank South recalled about the crucial minutes when the Rangers were trying to make the beach. "In coming in the water, a wave would hit the back of his head and knock his helmet off. So he'd hand his rifle to me and then he'd start walking, almost underwater, looking for his steel helmet again. This happened four or five times. That's the funny part about it. Not at the time."

The Germans manning the guns at the Pointe had been stunned to wake up and see the vast Allied flotilla in the English Channel. Overwhelmed by the magnitude of the invasion, they all, nevertheless, pulled themselves together to fight back. They weren't cowards. The Nazis engaged the Rangers with a fusillade of fire. They quickly tried to cut the Rangers' ropes without making themselves easy targets in the process.

Colonel Rudder, setting a fearless example for his men, was on the first LCA to land. He was followed by sixteen men from E Company, four from HQ Company, and his radio operator. He raced to the cliff and established a small headquarters in a cave. As the Rangers sprinted from LCA to cliff, a phalanx of bullets danced in front of them on the beach. Shingles went flying madly in all directions. (There was little or no sand on Pointe du Hoc beach.) Once at the cliff base, the equipment-heavy Rangers tugged on the dangling rope ladders to make sure the grapples had taken hold. "The tide was coming in real fast," Salva Maimone recalled. "So as we were getting up the cliff, the Germans were throwing grenades down the cliff—potato mashers are what they're called, concussion grenades—and different machine-gun fire from the beach flank. But we were so close to the Pointe du Hoc cliff, and under it, that we took good cover, so we started going up the cliff."

The Rangers lucky enough to have made it to Rudder's cave now took a deep breath, then started climbing. The difficulty—even the insanity—of their mission was more apparent than ever. Sergeant Eugene Elder, of F Company, had grown up in the Midwest, drifting between homes in Missouri, Iowa, and Kansas. Drafted into the Army in July 1941, Elder had trained in the scorching heat of the Mojave Desert before becoming a Ranger. Now, he commanded F Company's mortar squad. In an Eisenhower Center oral history, Elder recalls clinging to the cliffs, shouting out to his men, "Boys, keep your heads down, because headquarters has fouled up again, and it has issued the enemy live ammunition."

Scattered all around these huddled, cliff-clinging Rangers, pausing in the shadows, were grim early signs of deadly destruction. The Allied bombers that had tried to soften the area had knocked rock boulders off the side of the cliff. Avalanche was a new worry they hadn't really trained for. Everything looked wrong and eerie. Badly wounded men were yelling out in agony for help like in a scene from *The*

Red Badge of Courage. Some scholars, like Ambrose and Drez, claim that the Allies dropped 10 kilotons of high explosives on the Pointe du Hoc area—the rough equivalent of the "Little Boy" bomb dropped on Hiroshima. Pointe du Hoc hadn't just been "softened" by Allied bombing, it had been pummeled. "The enemy had time enough to get up out of those underground bunkers and shake his head, clear his brain and take us under fire," James Eikner recalled. "Fortunately there was a small flotilla of destroyers about and also the old battleship *Texas* had participated in the shelling prior to our landing. Of course, the Air Force and the Navy bombed, shelled, the Pointe du Hoc prior to our arriving but things had been timed so that the shelling and bombing would be letting up just as we were touching down at 6:30. Well, being forty minutes late you could see that the enemy was up and about."

A panoramic photograph of the beach in front of Pointe du Hoc on June 6 seems to show an apocalyptic wasteland or a moonscape on fire. Back in training at quaint Bude in Cornwall, the Rangers would not have been able to imagine such carnage. The Norman topography that they had studied on maps and models now seemed phony, futilely one-dimensional. Mayhem knew no longitudinal confines. The combination of loud booming noises, coming from all directions, added to the symphony of weird madness. Medic Frank South later recalled the difficult task he had of aiding the fallen first wave of D-Day invaders. "F company landed on the left flank of the landing area below the cliffs," South is quoted saying in Patrick K. O'Donnell's *Beyond Valor: World War II's Ranger and Airborne Veterans Reveal the*

Heart of Combat. "The LCA I was on got our grapnel rockets off in good order. But high up on our left flank was a machine gun nest that was in a superb enfilading position. We could not find it until later. We were caught in its field of fire. The LCA to our left came in so close that their rockets would have hit the side of the cliff instead of on top. Sergeant [John] Cripps dismounted the rockets from the boat and brought them in so they could be fired at a higher angle. This required that they be individually hand-fired using a 'hot-box.' Since he was under almost constant machine gun fire, this took enormous self-control and concentration, but Cripps was able to successfully put the grapnel hooks up over the lip of the cliff. As my LCA [884] landed, I still had the huge pack on my back to get ashore. I was aft in the landing craft—for good reason because we had to get the rockets fired, the lines up, and the climbers off first. We were under constant fire at this time, and my pack was so large it got in my way. As the others jumped off the ramp, the bow of the boat raised and the boat shifted slightly. When I jumped off I found myself in a shallow (at the time it felt very deep) shell hole, bent double so that my head was immersed. I realized that I well might be beamed by the ramp if it lifted in the swell and came down on me. I slipped the pack off and scrambled onto the beach. Reaching back, I managed to grab the monster and drag it in after me. The next thing I know someone yelled, 'Medic!'—a guy off to my left had been hit in the chest. At this time, a number of people were getting hit by small arms, machine gun fire, and grenades."

Back when the Rangers had been training at Camp Forrest in Tennessee, Rudder had preached the gospel of initiative;

that mantra now more than ever became Companies D, E, and F's guiding principle. Up against the cliff wall, looking directly up at the enemy rat-tat-tatting away with machine guns, they were forced into instant reaction mode. "The rockets with the grapnels attached were fired, bringing up the climbing lines and ropes," South reminisced. "Many fell short because they were fired too soon, or because the ropes were wet and heavy. The grapples were designed to go to the top of the cliffs, dig in there, and then we would be able to scale the cliffs, using either rope ladders, which some of them had, or simply straight ropes, which most of us preferred."

Another equipment setback the Rangers encountered was that the DUKWs were unable to land. With the high tide and cratered beach and relentless enemy fire, they were forced to stay in the Channel. The four DUKWs had been equipped with extension ladders from the London Fire Brigade. One had sunk on the run in and two of the others could not reach the narrow land. But the fourth edged forward and now lifted its ladder vertically and inched in toward the sky. "We had what the Army called DUKWs—they are amphibious jeeps," Corporal Louis Lisko of Pennsylvania, who was wounded and earned a Purple Heart that day, explained. "And there were fireman ladders mounted on these DUKWs, and there was a Ranger on the top of one, and there were extension ladders that you turned on the motor. This ladder started moving up to the top of the cliff, and this Ranger on the top had two British twin Lewis machine guns there. He was supposed to go up slowly and, if there were Germans on the edge of the cliff, he was supposed to fire on them. He had [an] armor[ed] panel underneath him to keep him from getting hit by bullets as the ladder was coming up. But, because of the bomb craters in the beach and in the water, the DUKWs were useless."

On top of the wobbly DUKW ladder was Staff Sergeant William Stivison, bobbing and weaving with the pitch and roll of the craft while manning the twin Lewis machine guns mounted on the top. As he came almost even with the level of the plateau, ninety feet in the air, he blazed away while the ladder swayed in midair. Stivison sprayed the top of Pointe du Hoc with fire while the Germans dived for cover. A few Nazis managed to fire back but were unable to kill the elusive Sergeant Stivison as he swayed back and forth, blasting away. Enemy tracers whizzed past Stivison—again, many came close, but none hit their mark. The aerial firefight lasted for a couple of minutes. Eventually, Stivison had to call it quits. Unable to win over the wind, Stivison had to be lowered back to the DUKW. His gallant attempt to eradicate the enemy had failed. So with no meaningful DUKW support, waterlogged ropes, time delays, and an entrenched enemy, the Rangers started to climb. "My thought was that this whole thing is a big mistake," Lieutenant George Kerchner recalled, "that none of us were ever going to get up that cliff." But the Boys of Pointe du Hoc, inch by inch, step by step, breath by breath, continued their ascent. How? A few LCA grappling hooks had penetrated the cliff—they were the Rangers' avenue to the top. "Rangers watched with sinking hearts as the grapnels arched in toward the cliff, only to fall short from the weight of the ropes," Ambrose wrote in *D-Day*. "Still, at least one grapnel and rope from each LCA made it; the grapnels grabbed the earth, and the dangling ropes provided a way to climb the cliff."

The plan had been for Companies E and F to land on the left side of the Pointe while Company D landed to the right. The delay in the approach, however, changed this strategy. First Sergeant Len Lomell had eyed an opening between the boats of E and F and ordered the remaining two surviving boats of Company D into the narrow gravel beach. The nine landing craft had a total of forty-eight mortar tubes to fire the grapnels at the top of the cliff. As each boat had come ashore, it had fired its grappling hooks, but only around twenty made it to the top. Still, it was enough. As the Rangers climbed, the Germans lobbed grenades down at them and tried to cut the ropes with knives. "The ramp goes down," Lomell recalled. "I'm the first guy shot, machine-gunned through the right side. I don't know if it was a machine-gun bullet or a rifle bullet. It just went through my side, through the muscle. And then I step off into water over my head. Believe me, I wasn't counting ropes. I came out of that water and I have my arms full of gear and stuff. The guys pulled me out; my platoon and I just rushed to the base of the cliff and grabbed any rope that we could get in our hands to get up that cliff. I couldn't tell you if they were my ropes or F Company's ropes or E Company's ropes."

A movie should be made on the life of Len Lomell. An adoptee, Lomell had a wonderful upbringing in Point Pleasant, New Jersey. Once he joined the Army in 1942, he rose quickly in the ranks. Of medium height, well-built, with brown hair, blue eyes, and a slightly hooked nose, he was, as the cliché goes, a natural born leader. Although he would

Sergeant Len Lomell, 1944. (Courtesy of the Eisenhower Center for American Studies, University of New Orleans)

vehemently deny it, perhaps more than any other 2nd Ranger Battalion soldier, his guts and cunning transformed a questionable outcome at Pointe du Hoc into an unequivocal victory. "It was all happenstance," he recalled in early 2005 from his home in Toms River, New Jersey. "We were at the right place at the right time to really damage the enemy."

Lomell rose to the occasion when Company D, having abandoned the right side, added the weight of their attack to the left side of the Pointe. Despite his bullet wound, the sergeant climbed the cliffs. He made it to the top in good time. Once up there, he gathered his men and rushed the German pillboxes and bunkers. As historian Ronald Drez has noted, Lomell and company, worried about snipers, collectively embodied that character in a Bill Mauldin cartoon who quips, "I feel like a fugitive from the law of averages."

Scampering from hole to hole like prairie dogs eluding rattlesnakes, the small Lomell-led Ranger force advanced toward three casemated German gun positions on the right side, near where they were originally to have ascended the cliffs. When they got to their position, instead of having to attack heavily defended gun emplacements, the Rangers stared into empty pits and casemates. Wooden logs—decoys—had replaced the 155-mm guns. Their luck had changed. The Germans, worried about air bombardment, had retreated and had taken their menacing guns with them about 500 meters away. "We didn't stop," Lomell recalled. "We played it just like a football game, charging hard and low. We went into the shell craters for protection, because there were snipers around and machine guns firing at us. We'd wait for a moment, and if the fire lifted, we were out of that crater and into the next one. We ran as fast as we could over to the gun positions—to take the one that we were assigned to. There were no guns in the positions." That's when Rudder's belief in Ranger initiative took over. Lomell continued, "So we decided, 'Well, they must have an alternate position somewhere.' And we thought, 'We'll hear them. Maybe we'll see some evidence of the movement.' But we never did hear them."

The resourceful Lomell did not waste time lamenting the empty casemates. Since his primary mission to destroy the guns was not attainable, he immediately turned to his secondary mission, which was to interdict the coastal road that ran behind the Pointe and joined the German positions all the way from Grandcamp to Sword Beach at Ouistreham. "We never stopped," Lomell recalled. "We kept firing and charging all the way through their buildings area, where they came out of their billets in states of undress. We were confronted with them there on our way up the road from the Pointe to the coast road. Our orders were to set up a roadblock and keep the Germans from going to Omaha Beach. We were to also destroy all communications visible along the coast road."

Lomell's eleven men from Company D were the first to reach the coastal road, which was blacktopped. They set up roadblocks both left and right, but they were hardly in position when the terrifying sound of Nazi marching boots was heard headed in their direction. Quickly, the small Ranger force ducked for cover as a forty- to fifty-man German patrol appeared heading toward Omaha. Peering out from shrubs, the far outnumbered Rangers let the Nazi soldiers pass. The Germans turned into a field and faded from sight, and the Rangers reset their roadblock. "They were a heavily armed group of Germans," Lomell recalled. "We hid and said, "Let'em go!"

Throughout Ranger training Lomell's best friend was Sergeant Jack Kuhn, who hailed from Pennsylvania. Fascinated with all things military since childhood, when he turned sixteen Kuhn signed up for the Citizens Military Training Camp, at Fort Meade, Maryland. A big career setback occurred when the Marines refused to enlist him due to colorblindness. He abandoned hopes of a military career and got a job working for an Altoona-based railroad company. But Pearl Harbor reawakened his desire to serve in uniform. In May 1942, he enlisted in the Army. Infatuated by the movie *Northwest Passage,* Kuhn made joining the elite

Jack Kuhn, 1944. (Courtesy of the Eisenhower Center for American Studies, University of New Orleans)

volunteer service his number one goal. Now, with Lomell at his side, he was liberating Europe. It wasn't a popcorn movie. They were playing for the highest stakes of all: life and death. "He was a real hard-charging Ranger," Lomell recalled. "Along with Harold 'Duke' Slater, he could really climb those ropes. We used to call them Batman and Robin."

So now Lomell and Kuhn, like Butch Cassidy and the Sundance Kid, found themselves not just up the cliffs but wandering around enemy territory. They trotted a short distance along the coast road to the right of the intersection. Before long they found an obscure, narrow country lane. They saw equipment tracks imprinted in the dirt: a clear sign that heavy machinery or loaded wagons had recently passed by. "And so Jack and I went down this sunken road not knowing where the hell it was going, but it was going inland,"

Lomell recalled, "and we came upon this vale, or this little draw, with camouflage all over it. Lo and behold, I peeked over this—just pure luck—over this hedgerow and there were the guns, all sitting in proper firing condition, the ammunition piled up neatly, everything at the ready. But they were pointed at Utah Beach. They weren't pointed at Omaha Beach."

Sergeant Lomell lay in the hedges for several moments looking at the five abandoned German guns next to an apple orchard, his mind frantically racing. There they were! The guns of Pointe du Hoc! Sitting unguarded in a field! He did, however, see a group of about one hundred Nazi soldiers in the far corner of the field, looking as if they were having an impromptu strategy meeting. Lomell, realizing that hesitation wasn't an option, saw his chance to wreak havoc, to destroy the five German guns that had been moved from Pointe du Hoc and were now at his reach. (The sixth gun had been decommissioned in a prior air raid.) "I said, 'Jack, you cover me, I'm going in there and destroy them.' So, all I had was two thermite grenades. I went in, he covered me. I said, 'Keep your eyes on these people. I won't know if anybody comes, and you keep your eyes open.' "

A paradoxically brazen but cautious Lomell crept from his hidden position in the tall hedgerow line and slipped into the field, one eye on the guns, the other on the German soldiers only seventy-five or one hundred yards away. With the steady hands of a safecracker, Lomell put one of his two thermite grenades into the elevating and traversing mechanisms of the camouflaged German gun closest to him. He then crawled to the second one and repeated the process. He pulled the pins and crept back to the hedgerow as the thermite grenades—which resembled soup cans—silently ignited. He had decommissioned two out of the five. "There were no craters around the apple orchard," Lomell recalled. "So we knew that the Germans, before we came around, had successfully hidden the guns."

Officially known as AN-M14 TH3 incendiary grenades, these thermite grenades were created so GIs could easily destroy enemy equipment. When placed on metal, as Lomell had done, these pyrotechnic devices generated 4000° F. of heat for thirty to forty-five seconds and burned straight through steel. The white-hot heat soon melted the metal of the mechanisms and rendered the guns immovable. They could still be fired, but only in their locked positions. The Germans never turned or moved from their position in the far corner of the field. "There's no noise to a thermite grenade that could be heard a hundred yards away," Lomell recalled. "These grenades were used especially for this sort of job because they melt gears in [the] mechanism. But then we ran back to the road, which was a hundred yards or so back, and got all the other thermite grenades from the remainder of our guys. So we stuffed our jackets and we rushed back, and we put the thermite grenades, as many as we could, in traversing mechanisms and elevation mechanisms and ganged the sites."

Lomell again left the field, the unaware Germans still conferring. Breathless, he joined Kuhn, who was diligently guarding the approaches. After rendering the five guns useless, the two Rangers started to leave the scene of the crime. But a gigantic explosion dropped them to their faces. Thinking it was a short round from the *Texas,* or the dreaded sixth gun, they scrambled to their feet and sprinted back to the road. Later they would learn that the earthshaking boom had been the destruction of a Nazi ammo dump by another Ranger patrol. After a rocky start, everything was working out in the Rangers' favor.

With the covert actions of Sergeants Lomell and Kuhn, the dreaded guns of Pointe du Hoc were now neutralized. Mission accomplished. It was not quite 8:30 A.M. The cliffs had been scaled, the road between Omaha and Utah beaches had been interdicted, the big guns had been put out of action, and Company D had been successful in warding off ferocious attacks from German infantry. The company had become the first American unit to achieve its D-Day objectives. For their actions, Lomell received the Distinguished Service Cross and Kuhn was awarded the Silver Star. Two of Rudder's Rangers had become legends. "We went out to the cliffs at Pointe du Hoc, and when we got a look at what they had been going through for three days, it was a seesaw battle," Ranger Jack Keating later recalled. "They'd push inland, and then get pushed back, almost into the sea again off the cliffs, and it was touch-and-go for three days. Our colonel, Colonel James E. Rudder from Texas, one of the greatest men that ever lived, military, was hit twice on D-Day and refused to be evacuated."

Officially the Battle of Pointe du Hoc was won on June 8, when the American flag was raised on the promontory and a bugler played sorrowful notes. A sad reality of the victorious Allied action was that while the Rangers were ultra-brave, they were not, like comic-book super heroes, bulletproof. They had endured a horrific 70 percent casualty rate. Yet, their mission had been a military success. The surviving Rangers spent the rest of June, for the most part, guarding captured German prisoners and burying their dead friends. Many then went on to fight in the Battle of the Bulge. Lomell, for example, was wounded three times after D-Day as the Allies pressed their way east into Germany.

One reason Rudder's Rangers received less widespread attention than they deserved in the decades following D-Day was Cornelius Ryan's nonfiction classic *The Longest Day*. Without question Ryan had written an enthralling page-turner, one that remains among the more riveting books about the Second World War. Unfortunately, Ryan blew the story about the 2nd Ranger Battalion at Pointe du Hoc. Because the six German 155-mm guns weren't found atop the Pointe by the cliff climbers, he surmised the Rangers' acrobatic derring-do on the fabled cliffs was, for the most part, all for nothing. "Ryan had called me for an interview after the war," Lomell recalled. "I told him I would only give it if he interviewed us Company D guys as a group. He refused. It was a mistake on my part. He completely botched the Pointe du Hoc story in *The Longest Day.*"

It took Ronald J. Drez, who was working on a fiftieth-anniversary volume titled *Voices of D-Day,* to locate both Len Lomell in New Jersey and Jack Kuhn in New York and get the story straight. "I still can't understand how Ryan blew it so badly," Drez said. "Those guns, with a fourteen-mile firing radius, were trained on Utah Beach. If Lomell and Kuhn, who became a police chief after the war, hadn't decommissioned them, all hell would have broken out and the U.S. death toll would have been much, much higher." Drez is correct. The point Ryan missed in *The Longest Day* was that if the Rangers hadn't climbed the towering cliffs by rope ladder, the German guns would have been quickly rolled back into firing position, ready to destroy the Allied armada. The Allied blood spilled at Utah Beach—nearly 200 men killed—would have been extraordinarily higher had those menacing guns not been permanently neutralized by Lomell's thermite grenades.

Although it took Drez to correct the history books, Lomell, back in 1984, hoped the showcasing of the Pointe du Hoc Ranger Monument, along with Reagan's speech, would finally eradicate Ryan's misguided proposition. Others felt the same way. "Cornelius Ryan pointed out that since the guns were not there perhaps the Ranger mission should not have taken place," James Eikner recalled. "That it would have saved a lot of lives if [it] had not. That conclusion of his wasn't really correct."

The official "after-battle report" for what transpired at Pointe du Hoc was written on July 22, 1944. Much of it is laden with military codes and arcane jargon, but some of the prose is riveting. "Ropes being wet, many rockets failed to carry over the cliff," the report, now archived at the U.S. Army Military History Institute at Carlisle, Pennsylvania, accurately summarized. "Men went up cliff by those ropes which had anchored and by scrambling under heavy MG and Sniper fire and a constant rain of grenades. Initial base established in cave at foot of cliff."

What had transpired during the Battle of Pointe du Hoc, and during the long, grueling months after the initial Normandy invasion, eventually *did* lead to the liberation of Europe. "General Eisenhower informs me that the forces of Germany have surrendered to the United Nations," President Truman announced in a V-E Day message at 9:00 A.M. on May 8, 1945. "The flags of freedom will fly all over Europe." Colonel Rudder declared V-E Day the greatest moment of his life—besides getting married.

Rudder's personal correspondence from 1944, housed at Cushing Memorial Library at Texas A&M, reveals what a truly agonizing year he had endured. Part of his responsibility as commanding officer was to write condolence letters to the families of Rangers killed in the European Theater. Here's one addressed to the mother of Corporal Willis Caperton, a Ranger who was killed in France:

Dear Mrs. Caperton:

No Commanding Officer can ever find words to adequately express his deep sympathy with those whose sons, husband or brothers finished their earthly tour of duty while under his command. The soldiers who formed the Ranger Battalion were the best—all volunteers. Their strenuous training and carefully supervised work and recreation brought officers and men, and the men and their buddies close, so close that they learned mutual respect and a dependence upon each other to carry out the individual tasks in order that the unit plan might work. The Day of Invasion found these men ready, fully trained, fully equipped not only physically but mentally. On the day before they came to France, men of all Faiths gathered with the Chaplain and dedicated the work at hand to God and their individual part in it as subject to His Holy Will.

The mission of the Rangers was successfully accomplished but as with all worthwhile things, the cost was great, so great indeed, that it cost the life you cherished and lost us a comrade and a friend. A Country must be great to call for the sacrifice of such men but America will always be great just because such men have fallen in order that the principles expressed in our Constitution might endure.

Every public honor will be accorded his memory. His President has already proclaimed him a hero. A grateful Congress will erect a monument to his name. The people of America will realize what that Gold Star means to those who loved him and will resolve to keep America worthy of such men. More than all these, however, the surviving Rangers, his buddies, will carry with them all their lives the example of his courage and will do their best to instill a like nobility in the hearts of generations to come.

So our comrade has gone and we realize that there is a void in your heart which neither your Country's gratitude nor our sympathy can fill. We, with whom he shared his life[,] ask only now to share his memory that it may inspire us all to the gaining of an early Victory and the making of a lasting peace.

With deepest sympathy,
James E. Rudder
Lt. Col. Infantry
Commanding

As for Colonel Rudder, who had been wounded three times, he returned to Texas after the war a hero, a modern-day Travis or Bowie. Proudly pinned on his chest were the Distinguished Service Cross, Silver Star, Bronze Star, Purple Heart, and French Légion d'Honneur with Croix de Guerre and Palm. While in Europe, besides missing his wife, he had daydreamed of being a rancher. Within a matter of a few years, he acquired a large cattle herd back in Texas. A Stetson hat replaced his Ranger helmet. As an active member of the Democratic Party, he became politically interested in helping the hamlet of Brady prosper. For six years—1946 to 1952—he served as mayor. After a brief spell as Texas land commissioner, he bowed out of politics to become a vice president of Texas A&M University, then president. A brouhaha ensued in 1963 when Rudder, after being president for five years, decided that the all-male Texas A&M should allow female students. "If Texas A&M goes on fighting and resisting this, we'll find ten years from now that we are still about the same size and the two major public universities in this state will be the University of Texas and Texas Tech," Rudder pronounced. "Now is the time to change and move on."

History proved Rudder right for pushing the Texas

A&M University system into the modern world. Unfortunately, after surviving both D-Day and the female enrollment controversy, Rudder suffered a cerebral hemorrhage and died on March 23, 1970. He was only fifty-nine. Anybody who knew him said it was way too young for a man of such indomitable character to die.

Ronald Reagan served in the U.S. Army Air Corps during World War II. He was based in California. (Courtesy of the Ronald Reagan Presidential Library)

4

REAGAN'S HOLLYWOOD WAR

The Rudder's Rangers who returned home in 1945 embraced the gospel of normalcy, claiming jobs at bustling factories, small farms, and family businesses. Others—like Len Lomell, who in 1951 received a law degree from Rutgers University—used the benefits of the Servicemen's Adjustment Act, the "GI Bill" that FDR had signed into law the previous year, to get a college education. But, perhaps because the Cold War came so quickly on the heels of World War II, forcing the United States to defend Berlin and intervene militarily on the Korean peninsula, there was no time for proper remembrance of D-Day on its tenth or twentieth anniversary. With the exception of Walter Cronkite's insightful 1964 interview with former president Dwight D. Eisenhower back in Normandy, which aired as a CBS special, D-Day was regarded by the general public in America as just another World War II turning point, like Midway, Guadalcanal, or the Bulge. No active president, in fact, had visited the D-Day battlefield sites while in office until Ronald Reagan paid homage in 1984. In 1964, Lyndon Johnson had *thought* about going to Normandy for the twentieth anniversary, but due to an overbooked campaign-year agenda, he didn't. Likewise, Richard Nixon *almost* attended the thirtieth-anniversary ceremonies in 1974, but the all-consuming Watergate imbroglio derailed his tentative plans.

Some of Rudder's Rangers did make private pilgrimages back to Pointe du Hoc, poking around the desultory rubble like everyday tourists. Up until Reagan became president, however, few paid any serious attention to their historical raid, although being able to boast that you had "climbed the cliffs" meant something at Rangers' Association meetings and Veterans of Foreign Wars halls. Five of the 2nd Battalion D-Day heroes were inducted into the Ranger Hall of Fame at Fort Benning, Georgia: Walter Block, William Clark, Jack Kuhn, Leonard Lomell, and James Rudder. Rangers like First Sergeant John Elderly and Lieutenant Elmer Vermeer never dreamed any of their cohorts would be officially memorialized in any specific way. But when mentioning the carnage that had occurred at Pointe du Hoc at reunions or on the telephone, the veterans of the 2nd Ranger Battalion rightly swelled with pride. They were proud of the job they had done. Their Ranger patch was the only accolade they needed—or wanted. Among themselves they proudly spoke of "liberating the hell out of the Pointe" but sadly remembered their fallen friends, like Sergeant William I. Mollohan Jr. and Private Rolland Revels. "We left the area together, but the sounds and smells of the battlefield still remain vivid in my mind," Lieutenant Vermeer recalled. "Yes, you can see the battlefield and hear the battlefield, but it's the smell of death . . . that really penetrates everything. You smell it as soon as the shells explode and the bullets fly, long before the dead bodies of your comrades start to decompose. It is a smell you'll never forget."

Given the fact that Ronald Reagan conjured up many anecdotes about the Second World War with such powerful emotive clarity, as if they came from the very depths of his soul, it's somewhat surprising to remember that he himself never saw combat or smelled death. He spent 1941 to 1945 in the safe haven of California. "Reagan's war stories are real to him; his war stories were his war, in part because he was acting a role as a soldier himself, off to war while at home," historian Garry Wills noted in *Reagan's America.* "And his hometown, Los Angeles, was a strange mixture of real and make-believe war." The only battlefield action Reagan saw was in weekly newsreels, propaganda shorts, and training films. His notion of D-Day Rangers came from Ryan's *The Longest Day,* with its questionable reporting of their actions on June 6. Yet he never tried to pretend otherwise. He never tried to claim glory for himself. Often, however, he would invent war stories—parables really—that made the stoic valor of GI Joe radiate even brighter than a *Stars and Stripes* propaganda report.

Never too concerned about factual accuracy, Reagan, for decades and in front of numerous audiences, told an apocryphal story about a B-17 pilot whose plane was crippled by enemy fire. They were spiraling downward, certain to crash. The pilot ordered his crew to bail at once, before they all died. Suddenly, the pilot realized that his belly gunner was bleeding profusely, too badly wounded to jump, wailing in fright about his imminent demise. "Never mind, son," the pilot shouted out, "we'll ride it down together."

The problem with Reagan's B-17 story is that if the pilot and gunner died together, who lived to tell the story? The answer is Reagan, who fancied himself from around Pearl Harbor onward as the raconteur of Ambrose's "we" generation. But, to his credit, he was self-deprecating about his own service. The most moving World War II story in Reagan's autobiography, *An American Life*, in fact, comically relates how his poor eyesight kept him from going overseas. "When my turn came for a physical exam, everything was fine except my vision without glasses," Reagan wrote. "One of the doctors who was administering the test told me after checking my eyes that if they sent me overseas, I'd shoot a general. The other said, 'Yes, and you'd miss him.' "

Instead of fighting fascism abroad, Reagan was placed on active duty in the United States on April 14, 1942. He was, as historian Edmund Morris wrote in *Dutch*, a "celluloid commando." But his interest in all things military could be traced back to his years growing up in Illinois and Iowa. He had first enrolled in a series of home-study Army Extension Courses as early as 1935. Dutifully he completed fourteen courses while working at various Hawkeye State radio stations, receiving "certificates of completion" in Military Discipline, Courtesies and Customs of the Service; Organization of the Cavalry; Military Law; Military Sanitation and First Aid; and Map Reading. With great excitement, he signed up with the Army Enlisted Reserve on April 29, 1937, as a private assigned to Troop B, 322nd Cavalry at Des Moines. Why the cavalry? Because Reagan had been enchanted since childhood by a passion for Zane Grey–like cowboys fading into an orange sunset riding their favorite horses with names like Trigger and Rifle. The very utterance of the word "cavalry" conjured up in Reagan's mind romantic Wild West notions of Cody and Custer on the gallop. Very soon, on May 25, Reagan was appointed second lieutenant in the Officers Reserve Corps of the cavalry. A few weeks later he accepted his officer's commission and was assigned to the 323rd Cavalry.

Reagan's first assignment during World War II was at Fort Mason, located between Fisherman's Wharf and the Golden Gate Bridge on lovely San Francisco Bay. The fort was the official San Francisco port of embarkation. Based around a huge pier and dock system, Fort Mason, the *New York Times Magazine* noted, dealt in a single commodity: "exporting war." Reagan's desk job entailed helping keep records on the 1.6 million troops and 23 million tons of cargo that were being shipped out of Fort Mason for the Pacific Theater. But, in truth, Reagan wasn't really a dockside body or bullet counter. He was too antsy and too famous—having made almost thirty motion pictures—to be ensconced behind a cold metal desk. From the moment he was drafted he assumed the role of an all-American Hollywood propagandist. Everywhere he went, people wanted to snap photographs of the actor who had made such popular films as *Brother Rat* and *Knute Rockne*. Even while he performed the dreary clerical tasks of a liaison officer, he signed autographs for men who were about to fight the Japanese on far-flung islands in the Pacific.

Luckily, San Francisco was not part of Reagan's life script for long. At movie mogul Jack Warner's urging, he applied for a transfer from the cavalry to the Army Air Corps in mid-May 1942; on June 9 it was approved. He was assigned first to Army Air Corps Public Relations and then, in a matter of days, to its First Motion Picture Unit. With colorful General Henry Harley "Hap" Arnold now his commanding officer, Reagan helped make Air Force training films, over three hundred of them.

The U.S. Army Air Corps was one in a series of many incarnations that American military aircraft units took on before the establishment of the U.S. Department of the Air Force in September 1947. During World War I what was then called the U.S. Army Signal Corps demonstrated military power from the air for the first time, conducting strategic bombing missions behind German lines. In 1926 this aviation sector of the U.S. Army was renamed the Army Air Corps, and upgraded into a branch of the Army. In June 1941 it was expanded once again, into the Army Air Forces, under which two organizations operated: the Army Air Corps for matériel and training (and, of course, training films) and the Air Force Combat Command for combat operations. But training films like the ones Reagan was making were a relatively new phenomenon. Under Arnold's direction, he would instruct pilots on everything from personal hygiene to surviving as a POW and avoiding the perils of booby traps.

By the end of World War II, Reagan had become a captain in the Army Air Corps. (Courtesy of the Ronald Reagan Presidential Library)

Reagan's Army Air Corps film unit essentially took over the Hal Roach Studios in booming Culver City, just ten miles down the road from Burbank and his old home lot, Warner Brothers. It was the ideal duty for Reagan. He could live at home in Pacific Palisades, make training films at "Fort Roach," wear a starched lieutenant's uniform, and feel good that he was serving his nation in wartime. And, surely, he was seldom bored.

According to Reagan biographer Lou Cannon, "Fort Roach" was known during this period for its "unmilitary wackiness." A random walk around the studio lot meant encountering a social ecology of everything from fake Japanese villages to makeshift Bavarian towns. With great fanfare Reagan starred in three well-known training films during the war—*Mr. Gardenia Jones* (1942), *Rear Gunner* (1943), and *For God and Country* (1944). (Others that he made were largely for internal Army Air Corps purposes.) Whether he was playing an Army chaplain, downed airplane pilot, or lonely soldier, Reagan proved to be a master propagandist, seemingly always speaking from the heart with bedrock conviction. Besides making these films, he occasionally briefed fighter pilots before they went overseas and traveled throughout America speaking at war-bond rallies and patriotic parades. Reagan never won an Oscar but everybody on Main Street sensed that the handsome Illinois native was pure red, white, and blue, a home-sired thoroughbred bursting with patriotic will. His rare gift, as political scientist Hugh Heclo explains, was an uncanny ability to offer up a "sacramental vision of America."

Today at the Reagan Library in Simi Valley, California, scholars can read over Reagan's 120-plus-page military record. For the most part it is standard fare. When asked during the Great Depression if he had any health ailments, Reagan wrote, on a couple of enlistment forms, "kicked by a horse, 1937." But on November 29, 1941—just eight days before Pearl Harbor—Reagan, then a second lieutenant in the cavalry reserve, listed a number of personal reasons that made active duty an impossibility:

a. I have four dependents that are totally depending on me for a living. The pay and allowances of a Second Lieutenant would be insufficient to meet their needs.
b. Due to the nature of my work, I do not feel that I should be ordered to Extended Active Duty with the

Regular Army until War is actually declared, at which time I will be more than glad to serve my country.

c. A recent report of physical examination taken at Fort MacArthur, California shows that I am permanently incapacitated for Active Duty due to compound myopic astigmatism, bilateral, severe, distant vision 6/200 both eyes, without glasses.

Besides training flicks and Main Street galas, the vision-deficient Reagan did perform in a major motion picture during World War II. It was composer Irving Berlin's blockbuster *This Is the Army,* made in Burbank. With his slicked-back hair, muscular physique, and pressed military uniform, Reagan presented himself as the ideal poster man for the U.S. Army. The message was clear: Uncle Sam had recruited Reagan—now he wanted you. While more than three hundred servicemen appeared in *This Is the Army,* Reagan was the genuine star. Relieved not to be making training flicks, Reagan performed admirably in the musical, which showcased such durable Berlin songs as "God Bless America," "We're on Our Way to France," and "This Is the Army, Mr. Jones." A highlight of the film was when the fifty-five-year-old Berlin, in a cameo, sang "Oh, How I Hate to Get Up in the Morning." (Heavyweight boxing champion Joe Louis and singer Kate Smith also made appearances.) Directed by Michael Curtiz, the Warner Brothers movie raised over two million dollars for Army relief. "As far as most of the cast, including Reagan, were concerned," historian Janice Anderson wrote, "the film was a military assignment."

During the filming, Reagan was shaken by an encounter

he had with Irving Berlin. Although they had met a few times before and Reagan was a Hollywood celebrity, Berlin acted like he didn't know him. At one point, after Reagan filmed a fine sequence with actors George Murphy and Alan Hale, Berlin pulled him aside for counseling. "You really should give this business some consideration when the war is over," Berlin suggested to him. "It's very possible you could have a career in show business." Have a career in it? Reagan had already made more than thirty movies. Always sensitive about his chosen craft and insulted by being ignored as a B actor, Reagan was wounded by Berlin's awkward attempt at a gracious remark. Had he made so little impact on the film industry that the legendary Berlin didn't know who he was? Was he just a forgettable matinee idol? Had he already become a has-been? Shouldn't he strive for something more?

It seems to me that Reagan's time in uniform during World War II transformed his self-awareness. (The same could be said about the other Cold War presidents who served as junior officers in World War II: Kennedy, Johnson, Nixon, and Ford.) By December 9, 1945, when Reagan was released from active duty, he had reached the rank of captain. This high-ranking military status meant a tremendous amount to him. But he could never become a genuine decorated war hero. A history buff, he didn't want to be just a B-grade—or even an A-level—actor. While he admired such talented colleagues as Clark Gable, Errol Flynn, and Gary Cooper, his true hero was President Franklin D. Roosevelt. Without question, as he evolved as a public figure, Reagan identified himself with the Squire of Hyde Park. Reagan, for example, considered his poor eyesight a handicap, an unwelcome ailment that had kept him out of combat. Roosevelt, however, had it worse: he was incapacitated by polio, trapped in the confines of his wooden wheelchair. Yet Roosevelt's fireside chats and inspiring wartime speeches, laden with determined inflections, were more powerful than twenty battalions of armed soldiers blasting away at the despicable enemy. "He gave confidence to the people," Reagan told Pulitzer Prize– winning historian David McCullough about his sky-high admiration for FDR. "He never lost faith in this country for *one minute.*"

Reagan came about his adulation for FDR via his father, Jack. A shoe salesman from Whiteside County, Illinois, Jack was a die-hard New Deal Democrat. No matter what the abbreviation—CCC, WPA, etc.—if it was part of Roosevelt's innovative alphabet soup, Jack Reagan was all for it. A good father, Jack always struggled to squeak out a decent living, moving his family around Illinois from town to town like a band of gypsies. Depending on the year or season, the Reagans lived in Tampico, Dixon, Chicago, Galesburg, or Monmouth. Never a good athlete, due in part to his poor eyesight, Reagan flourished as a thespian in college at Eureka State. His proudest accomplishment, however, occurred while working as a teenage lifeguard at Lowell Park, located along the Rock River in Dixon. From 1926 to 1933 he was credited with saving seventy-seven people from drowning. The town of Dixon, grateful for his heroics, erected a plaque in his honor. It was attached to a wooden log where each of his aquatic "saves" was acknowledged.

Like his entire generation, young Ronnie was also enchanted by radio. Determined to become an announcer, he took to imitating Franklin Roosevelt as early as 1932. That year he voted for FDR, and he would do so three more times. In 1933, when he began his radio career at WOC in Davenport, Iowa, he at times even sounded like the aristocratic, Harvard-trained president. His impersonation of FDR included chewing a cigarette holder, laughing loudly, and cocking his head, just like the President. "The field young Reagan had chosen was the arena of the political future already in this, his first year at the microphone," Garry Wills noted in *Reagan's America*. "Like others, many others, Reagan was soon mimicking the patrician diction that Roosevelt somehow made endearing with his penetrating tenor voice."

In 1935, Reagan moved to Des Moines, becoming a sports announcer at WHO Radio. While living at his boardinghouse, dreaming of eventually heading down famed Route 66 to Hollywood to make it big in the movie business, he got a glimpse of his all-seasons hero in the flesh. "The first president I ever saw was Franklin Delano Roosevelt from a distance of about thirty feet as he passed by in a parade in Des Moines, Iowa, in the 1936 campaign," he recollected in a recently published 1979 letter. "He was riding in an open car, a luxury I'm afraid presidents can't afford any longer in this day and age."

When Jack Reagan died in 1941, just before Pearl Harbor, Ronnie, who had recently played a Yankee in the Royal Air Force in *International Squadron*, was devastated. The father and son had, unfortunately, grown apart. But they shared that deep admiration for FDR. From Reagan's youthful perspective, FDR was *the* voice of democracy, a flamboyant wartime president blessed with an impeccable sense of theater. Reagan reveled in the high-octane radio addresses in which a fearless Roosevelt—sounding strong-willed and firmly defiant—attacked fascism while praising the enduring virtues of democracy and freedom. "There's no question about his leadership in that war as commander in chief," Reagan told newsman David Brinkley years later. "Our war effort was just absolutely magnificent, and we succeeded literally in saving the . . . world and probably achieved the greatest victory in . . . the history of war, in . . . the total surrender of . . . the provincial enemy."

Actors created throwaway personas, Reagan reasoned; FDR, by contrast, was a gallant persona for the ages. Playing George Gipp in a Hollywood film may have been a memorable role, but emulating Franklin Roosevelt in real life was an historic mission. When the President announced in his fireside chat on December 29, 1940, that the United States was the "great arsenal of democracy," Reagan was swept away. The splendid phrase stayed with him always. Later he memorized FDR's Four Freedoms speech of January 6, 1941, as if it were scripture. According to Lou Cannon, the Culver City cavalier who would one day enter national politics not only admired Roosevelt but "credited" him with saving American democracy. "It's clear that Reagan fashioned his own demeanor after the strong confident example of Roosevelt," author Mary Beth Brown noted in *Hand of Providence: The Strong and Quiet Faith of Ronald Reagan*, "and no doubt learned much from him about how a president should act when talking to a nation in difficult times."

After World War II Reagan continued promoting FDR's fierce internationalist crusade, even though his hero had died at Warm Springs, Georgia, on April 12, 1945. A registered Democrat who campaigned for Harry Truman in 1948, Reagan began seeing himself not as a mere actor but as the California tribune of FDR-style democracy. Roosevelt had prophetically warned his country during the 1930s of the gathering clouds of German and Japanese fascism. Pearl Harbor had proven him right. Now, even though the war was over, and "normalcy" was the buzzword of choice, Reagan continued to mount his Hollywood bully pulpit, talking about the "cost of freedom" and the "price of liberty" at county fairs, corporate events, and Sunday services. Like a one-man cheering section, he was keeping the rhetorical anti-Fascist flame of FDR alive. "I must say I think he was a great war leader," Reagan remarked to McCullough about Roosevelt. "I think there were less of the great tragic blunders that have characterized many wars in the past than this one. . . . I remember him when he . . . said that he was going to ask for 50,000 planes a year, and I remember when the American press tore him to ribbons for that. . . . They said that, you know, that this was impossible. It couldn't happen. But when you look at what this country did starting from the low point of Pearl Harbor in forty-four months—something like 350,000 planes and hundreds of thousands of tanks and/or trucks or every kind of weapon, we truly were the arsenal of democracy."

A turning point in Reagan's life occurred after he gave an anti-Fascist oration at the Hollywood Beverly Christian Church in the late 1940s. After Reagan's remarks the pastor approached the handsome actor with a suggestion: don't just deride Fascists, also add the imploding danger of global communism to your pulpit speech. Sure enough, he did just that the next time he spoke at the church. "If I ever find evidence that communism represents a threat to all that we believe in and stand for," he intoned, "I'll speak out just as harshly against communism as I have fascism."

This was the emergence of Reagan as committed anti-Communist orator. In the coming decades he proved an able student of the Rooseveltian tradition of monumental pro-democracy eloquence. He believed President Truman—a good and wise man—wasn't able fully to articulate the looming evils of global communism with the same intensity of purpose as his predecessor. It wasn't enough to sell Congress on the Marshall Plan and the Truman Doctrine—he had to sell to the people *directly* via radio and television, which was in its infancy. Sometime in the first phase of the Cold War, shortly before the Korean War, Reagan decided he would transform himself into a modern-day Paul Revere sounding the alarm about the threat of global communism. (It must be added, however, that Reagan believed American Communists should be allowed their constitutional right to speak their minds. He was not for censorship, nor did he approve of McCarthyism in any rabid way.) Over the coming decades, while the Cold War was in full swing, Reagan put his acting career on the back burner, preferring to prioritize politics. His micro-energies may have been consumed by his leadership roles in the Screen Actors Guild, but his macro-sense of destiny was pure Roosevelt. Just as FDR had spoken on radio *for* the American people, so would he. And the new menace, as he saw it, was the Soviet Union. Joseph Stalin, who had been chosen *Time* magazine's "Man of the Year" in 1939, was, in his mind, the new Hitler.

"Reagan believed that the world needed saving," Cannon explains. "He would answer the call."

After reading all the reliable books available on Reagan, particularly those by Dinesh D'Souza, Peggy Noonan, Michael Deaver, Garry Wills, and Lou Cannon, I'm convinced "Dutch," as he was called, never had as pronounced a political conversion from so-called FDR liberal to Goldwater conservative as some scholars would have us think. By the time Reagan switched from being a Democrat to a Republican in 1962, he had become an avatar of shrinking the federal government and believed John F. Kennedy was not hawkish enough in foreign affairs. Yet, even as Reagan became a Goldwater Republican he was always, in spirit and intellect, a Rooseveltian internationalist. Put more simply, as Reagan began his political ascendancy, he embraced FDR as commander in chief as his high-water benchmark. This was, however, quite distinct from Roosevelt as New Dealer, leading America out of the Great Depression like Moses. Reagan chose the wartime leader to emulate. His collected letters, published for the first time in 2003, reveal that Reagan often tried to justify decisions he made by clinging to the FDR leadership standard.

Never, at any time from the 1940s on, did Reagan abandon Roosevelt's optimistic, unwavering approach in foreign policy to confronting global menaces. At the 1980 Republican National Convention in Detroit, when Reagan accepted his party's nomination, he quoted Roosevelt over and over again. The next day, the lead editorial of the *New York Times* had the headline "Franklin Delano Reagan." While, as historian William Leuchtenberg pointed out in his *In the Shadow of FDR,* the *Times* delineated the political ramifications of Reagan's effort to "kidnap Franklin Roosevelt," it failed to note how much the two men shared. Both politicians held the unflappable belief that optimism was our national oxygen. As historian Frances FitzGerald noted in *Way Out There in the Blue: Reagan, Star Wars and the End of the Cold War* (2000), they were both "yes men" who, like Willy Loman in Arthur Miller's *Death of a Salesman,* embodied "a man way out there in the blue, riding on a smile and a shoeshine." Once he became president, Reagan, in what was essentially a weekly tribute to FDR, made upbeat radio broadcasts almost every Saturday. He had remained, all those decades later, an enduring fan of Roosevelt's fireside chats, even in the age of CNN and MTV. "You know," Reagan reflected in 1981, "it's pretty hard for most people to realize . . . at this time the impact of those fireside chats—that they, to this day, still hold the radio record for audience. When he came on, it was *the* biggest radio audience . . . *ever.*" And whenever Reagan was in a really tough situation, he asked himself a simple question: What would FDR have done?

The core of what Reagan learned from Roosevelt was that the world was divided into two camps: freedom versus totalitarianism. To Reagan's thinking, Stalin's USSR and Mao Zedong's China had become the new vortices of evil in the post-Hiroshima world. The great lesson of World War II, Reagan believed, was that the United States had to maintain its military superiority, and it had to stockpile more high-tech weapons than the Kremlin. Unpreparedness, at any level, Reagan believed, was unacceptable. As Reagan preached it, the nation's leaders in Washington, D.C., would have to keep the United States on high martial alert and enlist big business even more fully in the drive for a superdeterrence force. A permanent wartime economy, as had first been articulated back in 1950 by the National Security Council, had to be continued into the twenty-first century if America was to remain free from tyranny. By the time Reagan left the White House in January 1989, his foreign policy had been boiled down to three words: "peace through strength." Playing off of this maxim as president, Reagan used the Rooseveltian phrase "arsenal of democracy" over a dozen times, to plead for everything from the Strategic Defense Initiative (SDI) to aiding the Nicaraguan contras and upgrading the U.S. Navy.

Roosevelt, of course, had commissioned a bevy of fantastic speechwriters to help him get his emphatic message across during the Great Depression and World War II. Ray Moley penned FDR's memorable "forgotten man" campaign speech. Samuel I. Rosenman and Louis Howe collaborated on his immortal 1932 nomination acceptance, the "I pledge you, I pledge myself, to a new deal for the American people" speech. And Harry Hopkins and Robert Sherwood teamed up with Rosenman on the "arsenal of democracy" fireside chat.

The lesson was obvious: a modern-day president, to communicate effectively, needed poetic wordsmiths. The first official presidential speechwriter in U.S. history, Judson Welliver, was even called a "literary clerk." Starting in 1921,

Welliver toiled under the watchful eye of President Warren G. Harding. By today's standards, his handful of speech drafts seems meager. Every year since then, however, the statements and speeches delivered by the president—but written by somebody else—have increased in volume. Teleprompters have been employed and, at times, the president, like a nightly news anchor, is merely reading his script. Who composes the president's speech, therefore, exerts an enormous amount of influence over the political process. As Republican presidential nominee Thomas E. Dewey in 1948 famously quipped: "The man who writes the president's speeches runs the country."

But Dewey exaggerated. All of the powerful speeches quoted above *belonged* to Franklin Delano Roosevelt. His voice and attitude and determination and style—not the written words per se—are what soared into the history books. FDR may not have literally fought in World War II, but he was the statesman who personified the entire democratic crusade against Fascist evil. And when Reagan entered the White House on January 20, 1981, as the fortieth president, he assumed a similar statesmanlike role. The Cold War was still raging, and Reagan was in an offensive mood. Like Roosevelt, he believed expansive "liberty" declarations did change the minds and hearts of people everywhere. All he needed was a speechwriter who could make his voice ring like a bell. As fortune would have it, he found that person—not a man—three years into his presidency. Her name was Peggy Noonan.

Peggy Noonan with President Ronald Reagan, after she wrote the Boys of Pointe du Hoc speech. It was the first time they met. (Courtesy of the Ronald Reagan Presidential Library)

5

PEGGY NOONAN PREPARES FOR POINTE DU HOC

Growing up in Brooklyn, New York, Peggy Noonan dreamed of a career in literature or journalism. Born into a working-class family in 1950, she spent her childhood in various New York and New Jersey communities. As Noonan noted in *What I Saw at the Revolution* (1990), among her fondest memories, which would foreshadow her career trajectory, was living on top of a candy store in Rutherford, New Jersey. Eagerly she would purchase the *New York Daily News, New York Post,* and *Newark Star-Ledger,* among other newspapers, from the old-fashioned shop.

Determined to get a college degree, Noonan worked as a waitress in Rutherford and an insurance clerk in Newark to earn the tuition money. Eventually, in the fall of 1970, just months after the Kent State debacle, she enrolled in night classes at Fairleigh Dickinson University, which was only ten blocks from her home. "After two years I was accepted as a full-time student," Noonan recalled. "Fairleigh was another world. There were people there my age who did nothing but go to class and read, and they seemed so adult and experienced."

Her great love was literature, her major, and like a budding scholar, she devoured the novels of Fitzgerald and Hemingway. She was also intoxicated by the elegant prose of such cutting-edge new journalists as Tom Wolfe and Gay Talese. Before long she became editor of the school newspaper. Although the Vietnam War and American incursions into Cambodia were inciting students all over the country to protest, the economically strapped Noonan, while not entirely unsympathetic, just couldn't afford the luxury of dissent. "It was a hot, hot time and I would like to tell you that I wrote angry editorials and we took over the school, held the dean captive, and issued non-negotiable demands, but the fact is we didn't," Noonan wrote in her memoir. "It was New Jersey, and we were first-in-our-family college students, and we were working a job and studying and partying, and only rich kids wanted to occupy a dean's office, normal kids just wanted to not get called on the carpet there."

A key catalyst for Noonan's becoming a conservative was the first time she picked up a copy of the *National Review*, founded in 1955. Up until that moment she reasoned like a conservative but often felt alone in her beliefs. Like so many future Reagan White House hands, Noonan was aghast at the vociferous antiwar demonstrations held on college campuses and in District of Columbia memorial parks, where sniggering insults were hurled at the entire World War II generation. While she worried that U.S. Vietnam policy makers like Robert McNamara, the Bundy brothers, and Melvin Laird were dangerously fickle Dr. Strangeloves, and that both Johnson and Nixon had made a mess of things in Southeast Asia, she abhorred the romanticization of Marxist-Leninist-Maoist revolutionaries who denounced U.S. troops as baby killers. She was no budding Asian scholar but she knew Ho Chi Minh was not George Washington. "As far as I was concerned," she later wrote of the New Left, "they were encouraging the real bastards of the world."

Upon graduating cum laude from Fairleigh Dickinson, Noonan landed a job at the all-news CBS Boston radio affiliate WEEI. She worked the graveyard shift. Most important for later Reagan history, she began her career "writing for the ear." Putting her pen to good use, she carefully crafted editorials and commentary for various radio anchors to read on air. She was proud that the best media writers came from radio, not television. "I learned how to write for broadcast," she explains, "how to be conversational and catch the listener's attention, how to try to sum up a situation with a good line." Noonan received on-the-job training in Boston, studying the literary style of such legendary CBS broadcast writers as Eric Sevareid and Winston Burdett. "I listened closely to how they talked on the air, what they said, and how they said it," she noted. "I listened to headlines."

Eventually, her talent was recognized by CBS in New York. On September 7, 1977—her twenty-seventh birthday—she joined the legendary CBS team as writer/editor, working in the shadow of such Murrow boys as Douglas Edwards, Richard C. Hottelet, and Charles Osgood. The network gave her the midnight-to-8:00 A.M. slot. "It was my job to cut soundbites (in radio those five-to-twenty-second bursts of an interviewed person speaking were called actualities) for the *CBS World News Roundup,*" she wrote. "I also wrote the hourly news reports that I used to listen to in Boston."

By 1981 she was working for the indefatigable Dan Rather. Although Rather was deemed a liberal, one who had numerous public run-ins with Richard Nixon over the war in Southeast Asia, Noonan, the budding conservative, took a great liking to him. With his sharp Texas wit and tireless work ethic, Rather had risen to become the star newsman of CBS radio and television, the heir apparent to anchor Walter Cronkite. "Dan was a great boss," Noonan noted in a 2004 *Wall Street Journal* article. "He was appreciative of good work and sympathetic when it wasn't good. . . . He was open to ideas, he was democratic and not hierarchical in his management style, and he tried to be fair in his dealings with people in spite of a personal emotionalism that was deep, ever present and not entirely predictable."

Effervescent, poised, and attractive, Noonan had developed a smart, reliable writing style. Her words, regularly read by Rather on CBS Radio as his daily commentary, were refreshingly crisp and politically balanced. It was a four-minute essay that went out to all the CBS affiliates as "Dan Rather Reporting." She was clearly a star in the making as well as the token conservative in a predominantly liberal newsroom. "CBS, like all the networks, all media, was shaped in part by a certain political spirit," Noonan wrote. "It's a CBS tradition—Murrow on McCarthy, Cronkite on Vietnam, the correspondent of the thirties and forties warning what was happening in Germany. CBS drew people with a mission."

It was Kevin Lynch, then the articles editor at the *National Review,* who recognized the political change in quality of Rather's radio commentary once Noonan had entered the picture. A talent scout of sorts, he sought Noonan out. She sent him some of her radio scripts, which he read with genuine awe. A telephone call was made to Bentley Elliott, a Pennsylvania native who headed President Reagan's speechwriting department. He reported directly to Special Assistant to the President Richard Darman. Desperate for new talent, Ben Elliott was interested in meeting Noonan. An appointment was set up at the Old Executive Office Building (EOB) in Washington, D.C. "Kevin and I were friends," Elliott recalled about the hiring of Noonan. "We used to talk quite frequently about her talents. Kevin used to say that Noonan had managed to make Dan Rather sound conservative. That was quite a trick."

Noonan was thirty-three years old when she left New York and sublet an apartment in Washington. Rather's parting words to her were: "Don't let 'em scare you." Single, with dreams of someday having a family, Noonan was never shy of hard work. But she was not a careerist in the typical sense. She kept her priorities straight. Her faith in God was the most important thing in her life. A devoted Catholic, she tried to attend church services most Sundays, received communion, and occasionally went to confession. She was not, however, anything akin to an All Saints worshipper. Work, at this point in her life, was her altar. "Peggy's working-class Catholic background was a plus," Elliott recalled. "And her CBS scripts were terrific. I knew we had the right one." Whenever Noonan wanted literary inspiration, she turned to the King James Bible and read passages from Matthew, John, or Leviticus. "She embraces God fervently," Elliott said, "and helped the President articulate his own deep faith." But she also read urbane New York–based magazines like *Esquire* and *Vanity Fair*. In fact, it's fair to say that Noonan, the consummate wordsmith, liked the *language* of the Bible as much as its timeless religiosity. "One of our duties as speechwriters was to absorb everything that Reagan ever said," Elliott continued. "When you did that—like Peggy did—you realized that Reagan had a fundamental faith in God. He had learned it as a boy, inherited it from his mother."

Too often, historians—operating with a misunderstanding or bias—have painted the hawkish Reagan as a rank militarist. They point out that he shot down Libyan jets over the Gulf of Sidra, invaded the island-nation of Grenada, and taught the El Salvadoran government the dark ways of counterinsurgency. They also denounce his inflammatory "evil empire" language, which was aimed squarely at the Kremlin. This litany of complaints is, of course, true. But it is also accurate to note that Reagan, like Theodore Roosevelt and Dwight Eisenhower, worked overtime to keep America *out* of war. When on October 23, 1983, 241 U.S. servicemen, mainly Marines, were blown up in Beirut by a terrorist's truck bomb, a poised Reagan did not seek knee-jerk military revenge. A few months later, in fact, he withdrew American troops from Lebanon. These critics also overlook the fact that it was Reagan—not Kennedy, Johnson, Nixon, Ford, or Carter with their Soviet counterparts—who actually persuaded Mikhail Gorbachev, the Soviet premier, to seriously reduce nuclear weapons stockpiles at talks in Reykjavik and Geneva. Fearful of the New Testament's apocalyptic prophecy, Reagan believed his primary job as president was to avoid a cataclysmic nuclear showdown with the USSR. God, he believed, was guiding him. He saw himself as a peacemaker. Along with impassioned Rooseveltian oratory, a couple of his favorite quotes were John 14:27 ("My peace I give you") and John 3:16 ("For God so loved the world that He gave His only begotten Son, that whoever believes in Him should not perish but have everlasting life"). Clearly, Reagan's oft noted humbleness was, in the opinion of the people who knew him best, a direct result of his absolute faith in God. He lived, as much as one can, by the principles of the golden rule. "Reagan had a profound spiritual faith that grounded him and left him with nearly perpetual peace of mind," Mike Deaver explained. "This is not to say he didn't have his moments of doubt or anger, but those times were rare."

Noonan officially joined President Reagan's White House speechwriting team at the EOB in March 1984. She was thrilled to be working for *her* political hero. As one might expect, Noonan made immediate use of the White House library. She would plop down on a red velveteen chair in her cubbyhole of an office, turn on a reading lamp, and flip through large bound volumes of presidential speeches dating back to George Washington. She was hoping to garner ideas from the past to aid her in drafting future speeches. She admired the ringing oratory of Abraham Lincoln and Theodore Roosevelt, but it was Franklin D. Roosevelt who gave her a comeuppance. "All those head-wagging sentences, rounded, declarative, naturally written and naturally voiced," Noonan enthused in *What I Saw at the Revolution*. "You could see it on the page even if you hadn't read it all your life that the thing about Roosevelt was that he was shrewd and sunny, not hurt by man's sin but relishing its many varieties." She came to a realization: FDR, like Reagan, was comfortable in his own skin, a man who at the core believed "life is good."

As Noonan studied Roosevelt more closely she learned that he greatly admired William Ernest Henley's poem "Invictus," which concluded: "I am the master of my fate, / I am the captain of my soul."* It was clear to her well-trained ear that FDR had absorbed the poem's rhythm so fully it had metamorphosed into his own "triumphant cadence," a sound that "echoed not only in his speeches but in his recorded conversation." She recalled how Roosevelt would hold informal press conferences, a cigarette jutting out of his square jaw, twirling a globe around and explaining to the reporters why democracy was on the march in Argentina or

*Henley (1849–1903), an English poet, was handicapped by a bone disease. His verse was marked by an unchecked bravado.

Burma or Sicily. "I'd think," Noonan wrote, "This is how Reagan should sound." Her job was to make sure he did.

Noonan, however, was the new kid on the block of the White House speechwriting team. For her to get the plum assignments—and not get stuck writing Hallmark sentiments for First Lady Nancy Reagan—she would have to impress Dick Darman. A member of Harvard's Kennedy School faculty from 1977 to 1981, Darman had been one of eight or nine so-called assistants to the President when he joined the White House following Reagan's inauguration; he soon rose in both power and responsibility. He is credited with overseeing the initial Reagan economic program and its later modifications into the Tax Reform and Fiscal Responsibility Act of 1982 and the Social Security compromise of 1983. Scholars poring through the voluminous documents generated throughout Reagan's White House years will often see Darman's name in the top right-hand corner. He was the guardian, his cluttered desk the last stop before a policy speech, white paper, or postcard could reach the President's desk. By the spring of 1984, when Noonan arrived, Darman, whose father had fought in the Second World War, was overseeing approximately two-thirds of the White House staff; speechwriting was just one of his myriad responsibilities. While others covertly sniped at Noonan, deeming her a CBS liberal, a Ratherite, Darman embraced her robust talent. But that didn't mean he thought she was ready to write major policy statements like State of the Union addresses. He approved her hire to do *impressionistic* work, like funeral eulogies, anniversary remarks, and dinner toasts.

Under the tutelage of Darman and Elliott, Noonan's first

assignment was to write Reagan's announcement of the Teacher of the Year Award, a softball event if there ever was one. In *What I Saw at the Revolution* she related how speechwriters deemed crafting words for such fey events as "Rose Garden rubbish." Too new on the scene to complain, Noonan proceeded onward, penning Reagan's earnest remarks for that specific Rose Garden occasion, and others to come. They were well received. Given a thumbs-up from Darman, one of her next tasks was to write a speech for the President's upcoming visit to Normandy. He would be commemorating the fortieth anniversary of D-Day as part of a three-nation European tour and she would be providing some of his words for the grand occasion. "Reagan was very well read in history, particularly World War II," Elliott recalled. "You could look into Reagan's eyes and see fifty years of history, even one hundred and fifty years of history. Reagan's Pointe du Hoc speech, as it was gradually conceived, was for Reagan to unite the victorious lessons of our D-Day past with the contemporary challenges of facing down the Soviet Union."

Some of the old White House hands, touched with a degree of chauvinism, were piqued that Noonan had been assigned to work on such a potentially important speech. After all, she was a young woman who had never donned a military uniform, and a Washington novice to boot. Could she be entrusted with writing an important D-Day address? Did she understand the historical dimensions of what had transpired on bloody Omaha and Utah beaches? "It wasn't like New Jersey, where the housewives control life, or New York, where bright men and women rub shoulders and bump against each other a little too hard as they try to get ahead," Noonan recalled of her first White House days. "I had entered a place where men were completely in charge."

How Noonan went about writing the June 6, 1984, Pointe du Hoc speech can now be fully discerned by reading the files archived at the Reagan Presidential Library. As in all speechwriting, the first step in Noonan's laborious process consisted of gathering usable data about D-Day. A starting point for Noonan's research was acquiring a transcript of Reagan's June 6, 1982, remarks, which had been recorded on May 31 of that year in the White House for use on French television for commemoration of the thirty-eighth anniversary of the Norman invasion. For the most part the address read like a respectable Encyclopedia Britannica entry. It was full of hard facts and cold statistics but void of human drama. "D-Day was a success, and the Allies had breached Hitler's seawall," Reagan had intoned. "They swept into Europe, liberating towns and cities and countrysides, until the Axis powers were finally crushed. We remember D-Day because the French, British, Canadians, and Americans fought shoulder to shoulder for democracy and freedom—and won." Reagan did manage to cough up a positive remark about the "gallant" French leader Charles de Gaulle, who during World War II rose from colonel to leader of the Free French forces, but for the most part his address was pure boilerplate. No memorable headlines were made. No tearful personal stories of sacrifice were offered. No words honoring the World War II generation were espoused.

Clearly, for a thirty-four-year-old White House speechwriter brimming with talent and devoted to pithy *New Yorker*–style prose, such chronic dullness was unacceptable. Because the fortieth anniversary of D-Day was going to be an historic occasion, Noonan cast out a wide net, searching for inspired ideas from all corners. She received input from various foreign policy fiefdoms. J. L. Shub at the State Department, for example, provided a four-page memo laden with generic subheads like "Liberation," "Unity," and "Reconciliation with Adversaries." It was the kind of terse Foggy Bottom prose Secretary of State George Shultz would often deliver in a bland monotone, guaranteed to put a listless glaze over even the most ardent listeners' faces. The memo was bureaucratic, lifeless in a Harvard Law School type of way. Not a glimmer of the kind of high-note rhetorical bravado FDR would have demanded from his speechwriters Robert Sherwood or Samuel I. Rosenman could be found. It read like the deadly combination of a Ford drone and a dash of Carter tedium. The State Department also contributed a "country report" of France, which was, by its very mandate, cold, factual, and uninspiring. But at least Noonan had a standing start.

The well-intentioned State Department material hadn't inspired Noonan to develop a dramatic narrative. It did, however, instill in her the reality that Ronald Reagan's ten-day visit to Ireland, Great Britain, and France was much more than a drawn-out D-Day remembrance. In April, National Security Advisor Robert C. McFarlane, in fact, provided the President with a lengthy memorandum so he could "acquaint" himself at a preliminary stage "with the themes and objectives" of his upcoming European trip. Reagan's first stop would be Ireland, where he would speak at University College in Galway. As it turned out, Galway was celebrating its five hundredth anniversary as a port city and Reagan—whose father was a first-generation Irish American—would be anointed as an honorary citizen. He would receive the key to the city. And, even more impressive, Tipperary would name a pub after him and the O'Reagan graves in Ballyporeen would be scrubbed clean for his arrival. "Emphasize the economic and cultural ties and personal bonds between the President and other Irish Americans and Ireland," McFarlane's April 16 National Security Council memo instructed. It added that Reagan also needed to express the importance of "Americans to Ireland today through tourism, investment, technology flow, and job creation. Emphasize cultural ties with Irish authors, poets, and playwrights; songs and musicians and friendships between citizens." There would also be important meetings with Prime Minister Garret FitzGerald about the future of Northern Ireland and his assuming the presidency of the European Union on July 1.

The NSC memo basically claimed that because 1984 was an election year, the Reagans' trip to the Emerald Isle was laden with symbolism. "Although Reagan denied in an interview with Irish television that he was visiting his ancestral homeland for political purposes, his chief of staff, James A. Baker III, had no difficulty in acknowledging the political component," biographer Lou Cannon later recalled in a *Washington Post* article. "We have a bilateral relationship with an important ally and there are 40 million

Americans of Irish descent," Baker said. "Why should we apologize for this symbolism?"

As the itinerary dictated, after three nights in Ireland, Reagan would fly to London. Much of his highly anticipated talks with Prime Minister Margaret Thatcher would deal with economic questions like the U.S. budget deficit, unemployment levels in the OECD, and the international debt problem. The idea was for Reagan and Thatcher to spend June 5 together showcasing Anglo-American unity before crossing the Channel for the Normandy ceremony. On June 6, they would visit three spots in France: Pointe du Hoc, the American cemetery at Colleville-sur-Mer, and Utah Beach. "Normandy symbolizes the U.S. commitment to Europe, which led directly to the Atlantic Alliance," McFarlane's NSC briefing paper instructed. "The President will make brief (10–15 minutes) remarks at the Point [*sic*] du Hoc ceremony to about 5,000 people, including veteran groups. This should be emotional, stirring, and personal. The themes include reconciliation of former adversaries, how postwar cooperation has kept the peace for the longest period in modern European history, Alliance solidarity, and the strength of the American commitment to Europe."

What the NSC Report worried about the most, however, was President Reagan's not alienating West German chancellor Helmut Kohl, who had been banned from the Normandy ceremonies. In order that Kohl not feel too "bruised," the NSC wanted Reagan's speech to focus on reconciliation, with the announcement of a U.S.-Soviet summit on arms control to be the central message. In other words, there should be no Anglo-American gloating about defeating Germany forty long years ago. The Holocaust, the blitzkrieg, and Hitler were not supposed to be evoked, for fear such terms would make the speech come off as anti-German. With the Cold War dominating foreign affairs, the United States needed its NATO ally West Germany as a stalwart partner. Reagan's ripping the lid off the historical sewer of Nazism was deemed undiplomatic and strategically off-limits by both State and NSC. All the key White House people—including Treasury Secretary Donald Regan, Deputy Chief of Staff Mike Deaver, and Vice President George Bush—were sent a copy of the memorandum. Regan, a Marine officer in the Pacific Theater during World War II, became highly engaged in the importance of the Normandy event. "The State/NSC draft that I'd been given weeks before wanted the President to go on a little tangent about arms control, and as I read it I thought, in the language of the day, 'Oh, gag me with a spoon,' " Noonan bitterly recalled. " 'This isn't a speech about arms negotiations, you jackasses, this is a speech about splendor.' "

Given the NSC's preference for a speech about "reconciliation," Noonan, who was then unaware that so many D-Day veterans would be in attendance, was somewhat hamstrung. On the one hand she didn't want to repeat the thirty-eighth anniversary rote litany of facts. But she also didn't want to make the speech so pro-NATO—catering solely to the delicate Cold War disposition of Helmut Kohl—that it came off as flimsy Atlantic Alliance propaganda. An astute student of Reagan, she knew he was at his best when he told heartfelt stories about real Main Street people. That was his genius. Her boss was instinctive, blessed with a genuine show-biz gift for lively narrative and fabulist history. She didn't want to grind his address down just to please the men at State and NSC. As *New York Times* columnist Frank Rich rightfully noted, Noonan, in essence, was writing a "screenplay" for Reagan, not a mere speech.

As Noonan read Cornelius Ryan's *The Longest Day,* John Keegan's *Six Armies in Normandy,* and Jean Compagnon's *The Normandy Landings,* she realized anew just how unbelievably dramatic the D-Day invasion was. Somehow, she would have to cut through the bureaucratic thicket and find a way for Reagan, an old Culver City cavalry officer, to talk emotionally about the raw heroism of veterans with the uplifting conviction of FDR. It wasn't a difficult task. Susceptible to theatrics and imbued with a lifelong enthusiasm for symbolism, Noonan knew all the perfectly lined white crosses and Stars of David in the Colleville-sur-Mer cemetery would choke Reagan up. What she didn't know yet is that the story of the brave 2nd Ranger Battalion survivors—dozens of whom would be in attendance—would do so even more.

One of the early concerns Darman had about the Noonan speech had nothing to do with her prose and everything to do with how President Reagan pronounced Pointe du Hoc. "Should RR use Point du Hoc or Point du Hawk (Point de Hawk)," Darman wrote on White House stationery to Elliott. "Should he try the French pronunciation?" Deaver, worrying primarily about the morning TV audience, made the decision that Reagan's diction would be American—not French or Franglais. Phonetically he would say: Point (just like West Point) de Hawk (like the bird).

This way there would be almost no possibility that Reagan would flub his strongest lines.*

With the possible exception of Ted Sorensen during the Kennedy administration, speechwriters traditionally are somewhat *persona non grata* in the very White Houses they're working in (Nixon's working relationships with Pat Buchanan and William Safire were, perhaps to a more limited degree, other exceptions to the rule). Boiled down, the inherent problem speechwriters face is one of public perception: Americans want to think that their president is always speaking directly from the heart, that the high-minded rhetoric he utters at an award ceremony or college graduation or in the State of the Union address emanated from his own mind and pen. White House speechwriters, while absolutely necessary, are best kept out of sight and out of mind. As National Security Advisor McFarlane once put it: "Speechwriters aren't supposed to make policy." They are, in essence, viewed by cabinet-level officials as Oval Office mystique killers. Every time a speechwriter boasts to the press that he or she composed an important phrase or was the primary author of an intriguing paragraph his or her stock goes slightly up while the president's is diminished. Yet, as noted, they've evolved into being essential players in modern Washington life. "We speechwriters are translators not emphasizing Spanish into English but translating the bureaucratic into the poetic, the legalese into the elegant, the corporate into the conversational, the complex into the simple," one Reagan-Bush-era wordsmith explained.

*According to a White House staff memorandum of May 21, 1984, others who had input into the speech were James Baker and Ed Meese.

"The job of the writer is to zap the jargon with which academics or bureaucrats couch their purposes. . . . If a president were ever to deliver a televised address in the same form that some department functionary sent over to the White House, it wouldn't be a 'fireside chat,' because it would put out the fire."

When Noonan arrived in March 1984, for example, most of the other White House wordsmiths, her new colleagues, had not even *talked* with President Reagan in over a year. Not a word! They hadn't even shaken his hand when walking down a corridor. This was unacceptable to Noonan, who was used to having daily talks with Dan Rather while at CBS. Could Noonan institute a change? The protocol was to write the speech, give it to Elliott to edit, who would then pass it along to Darman for review. If Darman gave it a thumbs-up, he would place it in the President's in-box. Reagan would read the speech, make handwritten changes, and, if content, offer the comment "good job" or "well done" and initial the margin "RR." "All the speechwriters routed their scripts through me first," Elliott recalled. "That was the protocol. Once I approved, they would go to Richard Darman. He would circulate the speeches for critique. Eventually, the revised text would make it to the President's desk."

This hierarchical White House routing system, which was quite sensible, worked well. That doesn't, however, make it any less astonishing to realize that when Elliott assigned Noonan to write what would become the Boys of Pointe du Hoc speech, she was not dispatched to make an on-site inspection of the famed D-Day promontory. She had not, in fact, been formally introduced to Ronald Reagan himself. She had seen his feet through a door once, and proudly heard him read the remarks she had penned for a Rose Garden ceremony, but she had never actually *met* her boss. Not that she didn't lobby for the essential privilege of a proper introduction. "I can't write well without hearing the person I'm writing for talk in conversation," she protested to Elliott. "If you think he sounds stale, maybe it's because the speechwriters haven't met with him in more than a year!"

While Noonan—a new female speechwriter working in the back corridors of the EOB on an "impressionistic" Normandy address that Darman had commissioned—was an unknown to Reagan, the ubiquitous Deaver, who always had the President's ear, did the advance work for the upcoming European trip. A sure-footed showman, Deaver, whom journalist Lou Cannon deemed "the maestro of the D-Day production," actually credits his special assistant, William Henkel—a former Merrill Lynch corporate executive who had previously coordinated logistical arrangements for the 1982 economic summit at the Palace of Versailles—with traipsing the nine-mile-long Omaha–Utah beach area and selecting the visually stunning Pointe du Hoc as the location for Reagan's morning speech. Henkel's site assessment corresponded with that of John Vinocur of the *New York Times,* who wrote that May: "Pointe du Hoc is a knife stood on its edge, pointed into the sea. It looks lethal, a palisade of boulder and mean rocks where Normandy's green softness has reclaimed nothing. Battlefields: you could walk them from Gettysburg to Waterloo, and go back to your car, thinking of lunch. But not at Pointe du Hoc. The brightest morning roughens there, the wind working like a rasp, still scoring cruel edges on the sheer cliffs."

No matter what the day or hour or tide, standing on top of the craggy hundred-foot-high promontory known as Pointe du Hoc is an awesome experience. Gazing out across the Channel from this Norman watch post on a clear day, Deaver and Henkel strained to see Great Britain on the horizon. They failed, as all visitors do. It was too far away; nearly a hundred miles to be exact. Staring straight down from this vertical, fatal-looking outcrop, they did, however, ponder the slender strip of rocky, sandless beach and listened momentarily to the lulling surf. As they stood in the high-velocity wind, their suits flapping, they were both in awe. This was it, the exact spot where the U.S. Army 2nd Ranger Battalion made their fearless attack on June 6, 1944. As they wandered around, they were surprised some German concrete bunkers and blockhouses were still intact, that they had survived Allied bombing and nature's wrath. Rusted barbed-wire fencing was also still evident after forty years. They both knew, from a TV perspective, that this location—not Omaha Beach—would be the most dramatic backdrop for Reagan to speak. This intuitive decision was further validated by the fact that the Pointe du Hoc Ranger Monument—a dagger-shaped granite pylon—was going to be unveiled that afternoon overlooking the English Channel. It had been erected by the French atop a German concrete bunker to honor Rudder's Rangers. While it was not a *new* memorial, the ceremony would mark the transfer of custodianship to the American Battle Monuments Commission for perpetual care. Deaver, seizing on the simplistic beauty of the monument, already envisioned the Pointe du Hoc address being part of the Reagan bio film to be shown at the August 1984

Republican National Convention in Dallas. "I knew it would be our backdrop for the year," Deaver recalled. "Reagan's love of America and pride in World War II was just so *real*. He pined for that time, for those days that were gone. He'd say, 'You know, it used to be that if our country was in trouble, if a crisis was at hand, you just pinned a little American flag on your lapel and nobody harmed you. Nobody touched you.' "

Besides the breathtaking vista, both Deaver and Henkel were keenly aware—as Noonan wasn't—that approximately sixty Battle of Pointe du Hoc veterans would be in attendance to commemorate the fortieth anniversary (there wound up being sixty-two present). "I went over to France in advance with Henkel and others to set everything up," Deaver recalled in a November 2004 interview. "I was interested in the choreography of the event, making sure we had the right visuals. Remember, we had a fleet of ships off in the distance. We used the coast as a backdrop. All the TV producers were sent stories of 'the boys' in attendance in advance. As Reagan was saying those words—if you look at the film—the camera was going to flash on one or two of the Rangers. When Reagan acknowledged them, the camera would flash to a Ranger wiping away tears. It would be a very powerful image."

Deaver, however, had one hurdle to overcome if he wanted Reagan's speech to be delivered at 1:20 P.M. at the site of the Pointe du Hoc Ranger Monument: the French government. President François Mitterrand, the host of the D-Day ceremonies, insisted that Reagan meet for a photo op *before* he spoke at the Pointe. He wanted the ceremonies to take place later in the afternoon. But Deaver knew if he capitulated to Mitterrand's preferences, then his boss wouldn't be on the all-important U.S. morning TV shows. According to Cannon, Deaver put pressure on the French ambassador to the United States, Bernard Vernier-Palliez, to not make waves and to approve the 1:20 P.M. time slot. "The French moved up the ceremony," Cannon later wrote in the *Washington Post*, "to accommodate U.S. timing."

The Pentagon also was getting revved up for D-Day. On May 17 a ceremony was held in Arlington, Virginia, honoring General Maxwell Taylor and General J. Lawton Collins for their courageous participation in D-Day. Taylor had been the commanding general of the 101st Airborne Division, which was dropped behind German lines on D-Day while Collins had been commander of the U.S. VII Corps, which had successfully landed at Utah Beach. Also attending the Pentagon event was Wally Strobel, an Army paratrooper at Normandy. He was the 101st Airborne platoon leader from Kansas whom General Eisenhower was photographed instructing on the eve of D-Day. That image went all over the world. Now, forty years later, Strobel, the owner of Central Warehouse in Saginaw, Michigan, was ever so briefly back in the news. And, most important of all, on President Reagan's suggestion, the Pentagon unveiled a special D-Day exhibit. It included such artifacts as the postcard Rommel sent his wife after learning of the Normandy invasion, the flag flown by the U.S. Army Rangers after they had climbed Pointe du Hoc, and a signed copy of Franklin D. Roosevelt's D-Day prayer.

Secretary of the Army John Marsh was in charge of the

press event, but at an early juncture he turned the microphone over to Colonel Ray Stakes, a history professor at the University of Southern Mississippi in Hattiesburg and an active U.S. Army reservist, who was considered an expert on both D-Day and the Pacific War. His official military record noted that he was superb at writing "official documents and informal papers for historical accuracy."

Over the years Stakes had worked closely with the European Command, United States Army Europe, and the U.S. embassies in Paris, Rome, and Brussels. He knew his military history backward and forward. He began his lecture illuminating the quintessential roles of the 82nd and 101st Airborne Divisions on D-Day. He went on to talk about how Utah Beach was "the most unscenic area in Normandy." Then he elucidated, with great fanfare, the history of the Army's 2nd Ranger Battalion at Pointe du Hoc, which he colorfully described as "a spike that sticks into the English Channel." Because Reagan would be speaking on top of the hundred-foot promontory, Stakes went on, in clipped military fashion, to describe the Rangers' assault of June 6, 1944. "Three companies, of the 2nd Ranger Battalion, climbed those cliffs; landed by assault boat at the foot of those cliffs—in that area—put rope ladders up the cliffs, and assaulted the cliffs," he told the press at the briefing. "Their purpose was to neutralize gun positions, bunkers that the Germans had put on top of the bluff. When they got there, they found the bunkers were empty. They pushed on inland, discovered the guns that had been removed from the bunkers to back in a field, spiked them—destroyed them—and, at that time, were counterattacked by the Germans, and spent the next twenty-four hours literally fighting for their lives. They suffered. Of the roughly 200 men that assaulted the cliffs on the morning of June 6, only 75 of them came down when they were relieved a little more than twenty-four hours later."

The Reagan administration, at this juncture, was planning more than a dozen major World War II commemorative events to be held in Europe. The first would honor the June 2 liberation of Rome; Normandy would follow four days later. Reagan, as noted, was scheduled to deliver three separate speeches while in Normandy. But what was drawing advance media attention to Pointe du Hoc—besides that Reagan was scheduled to speak there—was the monument rededication and the fact that a reenactment of the risky June 6, 1944, assault by the 2nd Rangers was slated. U.S. Army veterans, most in their sixties at the time, were going to climb the steep cliffs. Once on top they would congregate around a German bunker that had survived Allied bombing. It had been built by Rommel when he was erecting defensive positions in 1943–1944. All around the returning veterans would be bomb craters, still ominous looking after forty years. The dagger-shaped stone memorial would then be unveiled at the ceremony hosted by President Reagan to honor the Rangers killed or wounded between June 6 and 8, 1944. The President would stand directly in front of the memorial, encircled by flowers. The plaque would read: TO THE HEROIC RANGER COMMANDOES D 2 RN E 2 RN F 2 RN OF THE 116TH INF WHO UNDER THE COMMAND OF COLONEL JAMES E. RUDDER OF THE FIRST AMERICAN DIVISION ATTACKED AND TOOK POSSESSION OF THE POINTE DU HOC. A French translation ran on the bottom half of the plaque. After spending less than an hour at Pointe du Hoc, the President would move on to nearby Omaha Beach, where he would deliver a major policy address.

Following the mid-May Pentagon press briefing, Noonan had a eureka moment. She comprehended fully for the first time the importance of the fact that surviving 2nd Ranger Battalion members would be sitting in the front rows when Reagan spoke at Pointe du Hoc. They would not be haphazardly scattered around the audience. They would be crunched together like choirboys in New England pews. As mentioned earlier, the White House advance team failed to tell Noonan that sixty-two proud veterans of the 2nd Ranger Battalion would be in the front rows to salute the President and hear his speech. Now, after the press briefing and an informal talk with Deaver, she also realized that a U.S. Ranger memorial would be unveiled. She scrapped her early drafts and started over. "There were some ways in which the Reagan speechwriting department was a little dysfunctional," Noonan told the *Atlanta Journal-Constitution* in June 2004, shortly after Reagan died. "One of the things they did wrong was send researchers, twenty-year-old kids, to the location of future speeches, along with the advance staff. The speechwriters were not sent. So a researcher went to Pointe du Hoc and she didn't come back with the kind of information that would be helpful to a writer because she wasn't a writer, and young people who aren't writers think writing is magic—you put the writer in a room like a plant and a pretty bud comes out. Anyway, I didn't know until shortly before the President left for Europe that the boys of Pointe du

Hoc—the old men who were the U.S. Rangers who took the cliffs of Normandy—would be there, in the first few rows, as RR spoke. I was indignant: How could you not tell me? RR will want to talk to them, not just talk over their heads! And thus, in the last days, 'These are the boys of Pointe du Hoc' was born."

While Deaver and Henkel were stagecrafting, and the Pentagon was in public relations mode, Noonan finished a preliminary draft of the Pointe du Hoc speech. This version, handed in at 1:30 P.M. on May 21, was reviewed by Ben Elliott, who marked it up and returned it to her. Drafts submitted in the coming days—May 23 and May 24, for example—were circulated to others for input. "What I learned from the Pointe du Hoc speech was that to the men of the Reagan White House, a good speech is really a sausage skin," Noonan later joked. "The stronger it is, the more you shove in."

As Noonan kept on reading, and talking to young administration officials who had done advance work in Normandy, she realized that Pointe du Hoc was going to be an ultradramatic historic spot for Reagan to speak. It wouldn't be his Gettysburg Address—as some foolish, history-deficient White House hands were already boasting—but she knew it *could* be a defining moment for Reagan's reelection campaign. And one suspects that a few of her lines might someday make it into *Bartlett's Familiar Quotations*. To her dismay, however, because neither Deaver nor Darman had ever sent her abroad to look over the promontory located directly between Omaha and Utah beaches, she was forced to rely heavily on published material. "I drifted around waiting for the speech to come; sometimes they do," Noonan later reflected. "I don't mean complete, I mean the basic shape and bits of literature of the speech start to present themselves. But the only thing that came to me was a phrase: *Here the West stood.* I tried writing on weekends, writing at odd hours."

Noonan started studying up on the surviving Rangers of the 2nd Battalion—and fast. The U.S. media was already promoting the fortieth anniversary of D-Day and it was still May. What gave Reagan's upcoming Normandy trip a real boon was *Time* magazine, whose May 28 cover story was "D-Day: Forty Years After the Great Crusade." Veteran journalist Lance Morrow did most of the analytical writing for the *Time* package, which was accompanied by a stunning Robert Capa photograph printed from his eleven surviving negatives of Omaha Beach. Morrow began his article with a quote from Shakespeare's *Henry V:*

> From this day to the ending of the world,
> . . . we in it shall be remembered,
> . . . we band of brothers;
> For he today that sheds his blood with me
> Shall be my brother.

Eight years later, in 1992, historian Stephen E. Ambrose would title his book about E Company, 506th Regiment, 101st Airborne, *Band of Brothers,* borrowing from Morrow, who had borrowed from Shakespeare. Ambrose also decided to use the dramatic Capa photograph of an American soldier wading ashore with the first wave of troops on Omaha

Beach, water up to his neck, determination on his face, which was the visual centerpiece for *Time,* as the jacket photo of his *D-Day, June 6, 1944: The Climactic Battle of World War II.* The packaging and repackaging of D-Day as a cottage industry had begun.

One can only marvel at how prescient and well-written the Morrow article was. Realizing that thousands of American World War II veterans were about to make a pilgrimage to Europe, Morrow imagined them wandering over "the pastoral killing ground," looking sadly for their fallen friends' names on simple white graves at the Colleville-sur-Mer cemetery. "They will remember exactly the spot where they were pinned down by German machine guns, or where a shell blast sent a truck pinwheeling," he wrote. "They will go up again to Pointe du Hoc and shake their heads again in wonder at the men who climbed that sheer cliff while Germans fired down straight into their faces. The veterans will take photographs. But the more vivid picture will be those fixed in their minds, the ragged, brutal images etched there on the day when they undertook to save European civilization."

Clearly Noonan had read the *Time* article. An underlined copy of it can be found in her Normandy files at the Reagan Library. And in her speech she quoted—as *Time* had done—General Matthew Ridgway lying on his cot, remembering God's promise to Joshua: "I will not fail thee or forsake thee." But what is even more significant about Morrow's piece is his trenchant analysis of why, in 1984, D-Day was about to become *the* election year symbol of the Reagan administration's New Patriotism. "The ceremonies in Normandy will celebrate the victory and mourn the dead,"

Due to the upcoming fortieth anniversary of D-Day, the American media started printing stories about GIs during the war. Pictured here are Army Rangers taking a break after knocking out a Nazi gun emplacement shortly after D-Day. (U.S. National Archives)

Morrow wrote. "They will also mourn the moral clarity that has been lost, a sense of common purpose that has all but evaporated. Never again, perhaps, would the Allies so handsomely collaborate. The invasion of Normandy was a thunderously heroic blow dealt to the evil empire. Never again, it may be, would war seem so unimpeachably right, so necessary and just. Never again, perhaps, would American power and morality so perfectly coincide."

Moral clarity. That was the ticket Reagan was trying to push to get reelected. What voter could argue that Adolf Hitler wasn't a villain worse than Idi Amin or Muammar Qaddafi? Who wasn't proud of the job America's Armed Forces had done in 1944–1945? According to the Reaganite view, NATO now faced an equally horrific threat from the Soviet Union. Munich-style appeasement was wrong in 1938, they believed, and it was wrong in 1984. But it was Morrow's understanding of how the D-Day story had spell-binding, redemptive qualities that Reagan could sell to Cold War America that really hit the mark. Morrow, perhaps placing himself into the President's mind-set or psyche, explained D-Day to *Time* readers as an American religious fable or sterling folklore moment. "The invasion, in a way, was a perfect expression of American capabilities: vast industrial energy and organizational know-how sent out into the world on an essentially knightly mission—the rescue of an entire continent in distress," he wrote. "There was an aspect of redemption in the drama, redemption in the Christian sense. The Old World, in centuries before, had tided westward to populate the New World. Now the New World came back out of the tide, literally, to redeem the Old. If there has sometimes been a messianic note in American foreign policy in postwar years, it derives in part from the Normandy configuration. America gave its begotten sons for the redemption of a fallen Europe, a Europe in the grip of a real Satan with a small mustache."

What the Reagan administration understood—as did Morrow—was that the American people craved something grander in their history and national memory than Gerald Ford's evacuation of Saigon or Jimmy Carter's malaise speech. Reporters used to write during 1979–1980, when fifty-three Americans were held for ransom by the Ayatollah Khomeini's Iran, that it was as if "America was being held hostage." By sharp contrast, the D-Day story was about America as ferocious liberators, not backroom barterers. Even though Reaganites tried to pretend for political purposes that the Vietnam War was a morally justified crusade, in their heads and hearts they knew better. Millions of Americans, and virtually every honest historian, recognized that the prolonged intervention in Southeast Asia was so rife with tragic political blunders that it was indeed an American failure. Wisely, Reaganites understood there was no winning way to build a consensus New Patriotism by reopening the controversial Vietnam wound. The United States had *wanted* to be D-Day-like liberators again in Vietnam, but that time around, for numerous murky geopolitical reasons, U.S. forces had become unwelcome invaders. That is why Reagan went all the way back to World War II—and Normandy in particular—to promote his New Patriotism during an election year. It was too hard to sell Vietnam triumphalism. But D-Day? That was a different story entirely.

Time wasn't the only periodical hyping the fortieth anniversary of D-Day that May. The *New York Post* ran a series of long, extended excerpts from historian John Keegan's *Six Armies in Normandy* and Max Hastings's *Overlord.* In late April the *Christian Science Monitor* had published a riveting story written by its World War II correspondent, Richard L. Strout. Likewise the May issue of *Smithsonian* ran a first-person account by Thomas L. Wolf, who had witnessed D-Day from the destroyer USS *Herndon,* firing off Utah Beach. "D-Day—the first time when the free world hurled its might, its treasure and the lives of its young men and women against the most powerful fortress ever erected: *Festung Europa,*" Wolf wrote. "Everybody knows the picture. This is the little picture." The *Milwaukee Sentinal* ran an essay entitled "Bloody Omaha" and the *Philadelphia Inquirer* had featured a story about intelligence breakdowns prior to D-Day.

But what was most noticeable about the pre-D-Day clips that Noonan collected as research were stories about the throngs of veterans returning to commemorate the fortieth anniversary. Because the Vietnam War had torn Americans apart for a decade, World War II veterans had been either marginalized or forgotten. There were, of course, in all fifty states granite memorials and reflecting pools honoring their sacrifice. But somehow the media had not focused on the uncommon valor of World War II fighting men since the tumultuous days when Ernie Pyle was firing off urgent dispatches from the trenches and Edward R. Murrow was boldly reporting on the radio from a bomb-besieged London. The American people *had* honored General Douglas MacArthur with a ticker-tape parade and General Dwight D. Eisenhower with a two-term presidency. But Reagan's election in 1980 had ushered in a new climate ripe for World War II remembrance. The New Patriotism was not just in the air, it was part of Reagan's DNA.

The media—following the Reagan administration's PR blitz—went along for the patriotic ride. Because Memorial Day is so close to the D-Day anniversary, almost every newspaper—large and small—suddenly decided to tell a story about the heroics of an unsung D-Day veteran. It was a twofer. The *Los Angeles Times,* for example, focused on a man named Clark Houghton while the *Miami Herald* chose Julius Eisner. Prominent in nearly all of these Memorial Day/D-Day tributes was a haunting reminder of the marble graves neatly lined up in rows at the American cemetery in France.

With the media on board, the D-Day anniversary drumbeat was clearly in full gear. Virginia senator John W. Warner even proposed that June 6 be officially declared D-Day National Remembrance Day. The thirty-six other members of the Senate who were World War II veterans immediately embraced his idea. Meanwhile, the press estimated that over ten thousand American veterans were planning to make a pilgrimage to Normandy to commemorate D-Day. "Now, with the fortieth anniversary of D-Day fast approaching, thousands who took part in the greatest amphibious and airborne operation in history are about to return to the scene of their epic deeds," the *Christian Science Monitor* reported. "Normandy, in short, is about to be re-invaded." Joining Reagan and the throngs of veterans would be a group of world dignitaries including French president François Mitterrand, British prime minister Margaret Thatcher, Canadian prime minister Pierre Trudeau, King Olav V of Norway, and Britain's Queen Elizabeth.

Because Pointe du Hoc had been chosen as the location for the first of the two principal D-Day commemorative speeches, Reagan approved the idea that the assault of the U.S. Army Rangers' 2nd Battalion be a central part of his address. With the right camera cutaways to teary-eyed survivors, Reagan could link his New Patriotism with the entire

World War II generation. As a longtime ardent admirer of the Rangers—and everything they stood for—Reagan wanted them to enter the national psyche as all-season heroes. The ball was now in Noonan's court to provide the linguistic magic—he was more than ready to step to the Pointe du Hoc podium and offer up a flawless performance.

President Ronald Reagan speaks at the Ranger memorial at Pointe du Hoc on June 6, 1984. (Courtesy of the Ronald Reagan Presidential Library)

6

REAGAN'S NORMANDY DAY

The more educated Peggy Noonan became about the 2nd Ranger Battalion on D-Day, the more impressed she was. Why didn't teenagers learn about them in high school? Why wasn't their dramatic "finest hour" story more widely known? She agreed with a statement Lord Mountbatten made in his foreword to historian James Ladd's *Commandos and Rangers of World War II:* "Today we are used to the daring exploits of 007, James Bond, but the story of these gallant raiders, commandos, Rangers and those who associated with them is even more exciting and gripping for these were real men facing real live dangers. It is time their story was told."

For the story to be told, it first had to be written. Noonan realized that the President's June 6 speech at Pointe du Hoc presented her with a rare opportunity to teach American TV viewers (the public)—albeit briefly—about the intrepid Rudder's Rangers, a unit that had disbanded on October 14, 1944, after racking up 18 Distinguished Service Crosses, 77 Silver Stars, 67 Bronze Stars, 585 Purple Hearts, and 2 British Military Medals. "The subject matter was one of those moments that really captures the romance of history," Noonan wrote in *What I Saw at the Revolution.* "I thought that if I could get at what impelled the Rangers to do what they did, I could use it to suggest what impels us each day as we live as a nation in the world. This would remind both us and our allies of what it is that holds us together."

The trick was to find two or three heart-wrenching D-Day stories from all these survivors to perhaps anchor Reagan's pep talk around. Her motive was straightforward: give Reagan anecdotal material that he could bring to life. With the right words, and such a dramatic Normandy setting, the audience would become a collective mist machine. "I wanted American teenagers to stop chewing their Rice Krispies for a minute," Noonan explained in her memoir, "and hear about the greatness of those tough kids who are now their grandfathers."

Noonan's book includes a chapter, "Speech! Speech!," that retells how she developed her Normandy address for Reagan. Her most famous line—"These are the boys of Pointe du Hoc"—came to her after reading Roger Kahn's bittersweet story about the Brooklyn Dodgers, *The Boys of Summer.* Kahn's elegant baseball book—his fourth—had received critical acclaim when it was first published in 1972. Because he grew up near the old Ebbets Field in Brooklyn, Kahn was able to provide engrossing first-person accounts of his beloved National League homeboys, who broke the color barrier with the phenomenal Jackie Robinson. But Kahn himself took "the boys" phrase from the Welsh bard Dylan Thomas's "I See the Boys of Summer."* (Noonan, oddly, never mentions the Thomas connection in her autobiography.) "O happy steal," Noonan later wrote of her borrowing of "the boys" phraseology from Kahn via Thomas.

Although some of Noonan's finest prose was edited out due to time considerations, a number of first-rate paragraphs made the final cut. For starters her opening: "We're here to mark that day in history when the Allied armies joined in battle to reclaim this continent to liberty. For four long years, much of Europe had been under a terrible shadow. Free nations had fallen, Jews cried out in the camps, millions cried out for liberation. Europe was enslaved, and the world prayed for its rescue. Here in Normandy the rescue began. Here the Allies stood and fought against tyranny in a giant undertaking unparalleled in human history." She had worked hard—against some NSC and State Department objections—to at least touch on the horrors of the Holocaust, if not mention it outright, in Reagan's address. She considered getting "Jews cried out in the camps" into the final version a coup. If there was one central NSC and State Department concern about

*The opening stanza of Dylan Thomas's poem is: "I see the boys of summer in their ruin / Lay the gold tithings barren / Setting no store by harvest, freeze the soils / There in their heat the winter floods / Of frozen loves they fetch their girls / And drown the cargoed apples in their tides."

Reagan's Normandy speeches, one that persisted even as he took the podium at Pointe du Hoc, it was about *not* offending our Cold War ally West Germany. Gingerly, Noonan avoided that pitfall without relinquishing her ardent position about evoking the death-camp victims in the speech.

Finding just the right prose rhythm, Noonan then penned the following evocative and descriptive paragraph:

> We stand on a lonely, windswept point on the northern shore of France. The air is soft, but forty years ago at this moment, the air was dense with smoke and the cries of men, and the air was filled with the crack of rifle fire and the roar of cannon. At dawn, on the morning of the sixth of June, 1944, 225 Rangers* jumped off the British landing craft and ran to the bottom of these cliffs. Their mission was one of the most difficult and daring of the invasion: to climb these sheer and desolate cliffs and take out the enemy guns. The Allies had been told that some of the mightiest of these guns were here and they would be trained on the beaches to stop the Allied advance.

Having set the scene, now she could deliver the heroic action—and the tragedy of death.

> The Rangers looked up and saw the enemy soldiers—the edge of the cliffs shooting down at them

*Noonan made a factual mistake. Because LCA-860 sunk with 35 men in it, and LCA-914 was disabled, about 180 actually jumped off the British landing craft and ran to the bottom of the cliff.

> with machine guns and throwing grenades. And the American Rangers began to climb. They shot rope ladders over the face of these cliffs and began to pull themselves up. When one Ranger fell, another would take his place. When one rope was cut, a Ranger would grab another and begin his climb again. They climbed, shot back, and held their footing. Soon, one by one, the Rangers pulled themselves over the top, and in seizing the firm land at the top of these cliffs, they began to seize back the continent of Europe. Two hundred and twenty-five came here. After two days of fighting, only ninety could still bear arms.

Besides honoring the Rangers, Noonan made sure other D-Day heroes were evoked. She added a reference to the Royal Scots Fusiliers, who battled to the beat of bagpipes at Sword Beach, and the Canadians at Juno Beach, who had overcome reefs, shoals, seaweed, and blazing German guns. After a round of inner White House debate, she even evoked, against her better judgment, the huge role the Soviet Union played in defeating Hitler's army. Robert Kimmitt of the National Security Council prevailed in this regard. "I have *not* incorporated this suggestion because it is irrelevant (the subject here is Normandy, and the Russians weren't at that party)," Noonan wrote Elliott on May 30, "unneeded (brings up the whole new topic of what losses each nation suffered in the war when we don't talk about the millions of French, British, German and American dead), and . . . it has that egregious sort of special pleading ring that just stops the *flow.* It sounds like we stopped the speech dead to throw a fish to the bear."

The final, agreed-upon version read, "It's fitting to remember here the great losses also suffered by the Russian people during World War II. Twenty million perished, a terrible price that testifies to all the world the necessity of avoiding war." It was, in essence, tossing the Kremlin a bone. But it was also a coy setting up of a Soviet straw man just to bulldoze it down. For Reagan's gracious recognition of the Russian contribution to the Second World War was accompanied by sledgehammer lines, pointed Cold War barbs hurled down in Zeus-like fashion in the middle of the Pointe du Hoc commemoration. "Soviet troops that came to the center of this continent did not leave when peace came," he complained. "They're still there, uninvited, unwanted, unyielding, almost forty years after the war. Because of this, Allied forces still stand on this continent. Today, as forty years ago, our armies are here for only one purpose—to protect and defend democracy. The only territories we hold are memorials like this one and the graveyards where our heroes rest."

Decades later, what impressed Ben Elliott about Noonan's finely crafted speech, besides, of course, the "boys of Pointe du Hoc" motif, was the tough, simmering, anti-Soviet Reaganesque language, which Kimmitt had provided. Impressionistic Noonanesque sentiments aside, the speech didn't give an inch when it concerned rolling back communism in Eastern Europe. It was the kind of feisty oratory American conservatives and British Tories relished. "Peggy's speech was a clarion call to the West to rally against the Soviet Union," Elliott recalled. "People forget that crucial aspect of the Pointe du Hoc speech because it's been remembered in history as only a great tribute to the World War II generation."

Besides World War II history, Noonan had found inspiration for the speech from a surprising source: the distinguished poetry of Stephen Spender. During the 1930s Spender had helped revolutionize British poetry. A hallmark of his verse was its sympathy for the values of Western civilization and a fierce sympathy for the world's dispossessed. Like an eager Ph.D. student, Noonan underlined verse from "I Think Continually of Those Who Are Truly Great" (a celebration of heroes) and "The Express" (a lament about modernity). Lines from the poem "Not Palaces" seemed, by the emphatic markings, to really grab her: "Our program like this, yet opposite / Death to the killers, bringing light to life." But Noonan eventually chose the following Spender verse to inject into the first third of Reagan's speech: "The names of those who in their lives fought for life, / Who wore at their hearts the fire's centre / Borne of the sun, they travelled a short while toward the sun / And left the vivid air signed with their honour." Part of this verse made it into the final draft.

Unfortunately, a few of Noonan's best lines hit the cutting-room floor. One omission, in particular, found in her original May 21 draft at the Reagan Library, should have stayed. It was designed to stir up a self-consciousness in the listeners, forcing them to hear the reverent silence of the commemorative moment. "As we stand here today," she wrote for Reagan, "the air is soft and full of sunlight, and if we pause and listen we will *hear* the snap of the flags and the click of cameras and the gentle murmur of people come to visit a place of great sanctity and meaning."

Taken together, Noonan had written a series of strong paragraphs, perfectly conceived both for Reagan's emphatic voice and earnest demeanor. But while everybody liked the opening paragraphs of Noonan's speech—and loved "the boys" bit in the middle—the end proved problematic. A frustrated Noonan had to slug it out with National Security Advisor Bud McFarlane, of all people, in order to end the Pointe du Hoc speech the way she had desired. She had wanted "borne by their memory"; he insisted on "sustained by their sacrifice." They argued in the typical bureaucratic way: as Noonan complained, "He'd x my phrase out, I'd x his phrase out. I told Ben: Look, this guy who talks like a computer, who in fact probably *is* a computer—I'm going to interface with him and tell him to leave my work alone. Ben told me: He's a computer with more power than you and he's the computer who'll be on the plane, and we'll all try very hard to preserve the integrity of your work." Luckily for Noonan, her guardian angel, Dick Darman, stepped into the fray and saw to it that at least regarding this particular speech, she prevailed.

Noonan wasn't the only White House speechwriter working on the upcoming Normandy events. While Noonan was the newcomer to the Reagan speechwriting team, Anthony Dolan, at the time only thirty-six years old, was deemed the old reliable hand. While he was good at impressionistic prose—for instance, the eulogy he had drafted for Reagan when Egyptian president Anwar Sadat was assassinated in 1981—it was not his forte; substantive foreign policy analysis was his bread and butter. Dolan grew up in Fairfield, Connecticut. His parents were Democrats disgruntled about the growth of international communism—Reagan Democrats before their time. When Dolan was thirteen or fourteen, he got involved with the burgeoning new conservative movement. At Yale University, where he majored in history and philosophy, he fell under the erudite spell of William F. Buckley Jr. It wasn't long before the *National Review* became Dolan's periodical of choice, and Barry Goldwater his adored candidate for president in 1964. Throughout the late 1960s and the 1970s Dolan moved between first-rate journalism and GOP politics. As an investigative reporter for the *Stamford* (Connecticut) *Advocate,* he exposed a classic nexus of organized crime and municipal corruption, winning the Pulitzer Prize in 1978 for his unflagging effort. Yet he worked overtime in 1980 as a campaign speechwriter, trying to get Ronald Reagan elected president.

Because Peggy Noonan had so effectively marketed herself in the 1990s and beyond as Reagan's official wordsmith, the speechwriting efforts of Dolan have been largely ignored. That is an historical oversight in need of remedying. An ardent anti-Communist, it was Dolan, not Noonan, who wrote Reagan's most memorable early phrases; for example, "evil empire," for better or for worse, was the product of his pen. Dolan was the only member of Reagan's senior staff to serve all eight years at the White House. (He was the chief speechwriter, special assistant, and deputy assistant to the President.) It was Dolan who crafted—in anticipation of victory—Reagan's 1980 election night oration after beating President Jimmy Carter. He ended up working on all eight of the President's State of the Union addresses. Straight out of the gates Dolan became, as much as anybody could, the

Speechwriter Anthony Dolan with President Reagan at the White House. (Courtesy of the Ronald Reagan Presidential Library)

foreign policy voice behind Reagan. When one considers he supervised the speeches for four Reagan-Gorbachev summits, penned the 1983 "evil empire" speech delivered in Orlando, Florida, and concocted the "crusade for freedom" motto for his boss to deliver to the British Parliament at Westminster Hall, it's amazing he has managed to stay so fully off history's radar screen. (As of March 2005, he was serving as Secretary of Defense Donald Rumsfeld's special advisor at the Pentagon.)

Part of the reason Dolan has been elusive is that while he had a fruitful working relationship with Reagan, he often fought with important White House players like James Baker, David Gergen, and Richard Darman. He also shunned the spotlight, preferring a career as a "confidence man" working in the top echelons of government. Also, because in the mid-1980s Noonan ended up capturing the public's imagination by writing two landmark speeches—"The Boys of Pointe du Hoc" and "The Address on the Space Shuttle Challenger"—Dolan was perceived by the media as a second-tier scribe. But it was Dolan, not Noonan, who started the tradition of Reagan's honoring men who served in uniform both past and present. "From the get-go I was inserting notions of conservative-sacrifice heroics into Reagan's speeches," Dolan recalled. "Great speeches are based on research. And I would look into stories about individual military heroism for Reagan. We would then honor in words, in major addresses, men who had won Congressional Medals of Honor or Bronze Stars."

A vivid illustration of Dolan's approach can be found in Reagan's January 26, 1982, State of the Union address. After evoking George Washington, Winston Churchill, Franklin Roosevelt, Douglas MacArthur, and John F. Kennedy—heavy historical hitters—Reagan talked about the more recent valor of Vietnam veterans. "We don't have to turn to our history books for heroes," he intoned. "They're all around us. One who sits among you here tonight epitomized that heroism at the end of the longest imprisonment ever inflicted on men of our Armed Forces. Who will ever forget that night when we waited for television to bring us the scene of that first plane landing at Clark Field in the Philippines, bringing our POWs home? The plane door opened and Jeremiah Denton* came slowly down the ramp. He caught sight

*Jeremiah Denton was a Vietnam POW, known for having blinked the word "torture" in Morse code throughout a forced interrogation by the North Vietnamese. He went on to serve as a senator from Alabama.

of our flag, saluted it, said 'God Bless America,' and then thanked us for bringing him home."

In the spring of 1984 Dolan was assigned to write a series of major foreign policy speeches for Reagan. The two most memorable would be delivered at Dublin and Omaha Beach. With Darman's approval he started working on both. (Dublin, in fact, was a big deal, because Reagan was going to proclaim he was open to holding nuclear arms negotiations with the Soviet Union.) Because of Dolan's Omaha Beach speech, Pointe du Hoc was considered a sideshow on June 6, a feel-good, non-policy-based moment. Therefore, Noonan was tasked with penning the "impressionistic" oration, while Dolan's "realist" policy speech was the one that the *Wall Street Journal, New York Times, USA Today,* and *Washington Post* would headline the day after it was delivered. Some in the White House simply referred to Omaha Beach as "the Speech" and Pointe du Hoc as "Brief Remarks."

Because President Reagan's impending trip to Normandy was getting so much advanced media attention, many families of D-Day veterans took to writing the White House. Most of them were congenial letters of gratitude, simple thanks to the Reagan administration for caring about what had transpired on Omaha and Utah beaches. "I started sifting through the World War II correspondence as it would come in," Dolan recalled. "My eyes were looking for something special. There was a deep synergy between Reagan and all of his writers. I think at a very elemental level we knew in the end it was his ideas and idioms that made our drafts work. And honoring real veterans was one of his core principles. He truly thought American World War II veterans epitomized patriotism."

More than perhaps any other U.S. president, Reagan loved both receiving mail from everyday people and replying to them. Back in the 1950s, when traveling all over America for General Electric, he frequently fired off missives to fans. A daily reader of magazines and newspapers, Reagan made it a lifelong habit to clip out articles he found interesting. What appealed to him most were stories about the vicissitudes of daily life found in magazines like *Reader's Digest* and *Look*. And he adored inspirational aphorisms. On note cards he would dutifully record pithy phrases that caught his fancy, words once uttered by Franklin Roosevelt, Plutarch, Alexander Hamilton, and others. Sometimes these venerable quotations would end up in his correspondence and speeches. Another Reagan habit was dusting off Karl Marx, his all-purpose nemesis. A gleeful anti-Communist, he would quote from *Das Kapital* just to slam Marx into the trash can of history. Reagan's genius as a correspondent was that he rejected elitism of any kind. Reagan didn't just write to high-profile people; he *listened* to middle-class Americans, answered their letters, took their questions and suggestions seriously.

A case in point was Lorraine Makler Wagner, who was working for the Internal Revenue Service in Philadelphia in the spring of 1984. Reagan's correspondence with Lorraine had begun in early 1943, when she was only thirteen. A devotee of such Hollywood fan magazines as *Photoplay* and *Modern Screen,* she sent Reagan a fan letter; to her surprise and delight, Reagan replied with a five-by-seven autographed photo, and then a penny postcard signed "Ronald," thanking her for her "swell letter." The correspondence between the two continued regularly over the following half century as Reagan went from matinee idol to Screen Actors Guild president, host of *General Electric Theater*, governor, and eventually president. The 276 heartfelt letters they exchanged between 1943 and 1992 document Reagan's political evolution with precision. After his first month in the White House, Reagan wrote his pen pal with a mix of personal and political feeling that is typical of their correspondence: "We're a little homesick for California," the fortieth president confessed. "But I must admit that I'm enjoying the opportunity to start working on the problems we have been thinking and talking about for so long."

Lorraine Wagner had reached out to him and he had genuinely embraced the episolatory gesture. He did the exact same thing with countless other "average" people, although not with quite the same frequency or duration. Realizing Reagan's penchant for reading fan mail—particularly those imbued with sentimentality—Dolan knew he had a winner when in March 1984 a woman from Millbrae, California, wrote the President searching for logistical advice on how her family could attend the fortieth anniversary of D-Day activities. Impressed by her story, Dolan sent the letter to Colonel M. P. Caulfield, deputy director of the White House Military Office, who was also smitten with this young woman's earnest, heartfelt prose. Without hesitation Dolan seized the moment and padded his Omaha Beach draft with parts of the Millbrae woman's adoring letter about her father, a deceased D-Day veteran. "I'll never forget the moment," Dolan recalled. "It was 2:00 A.M. and I was writing the Dublin speech. I stumbled upon this incredible letter. I had grown up around veterans' families and her voice was so real to me. And, therefore, I knew it would also be real to Reagan."

The woman, a flight attendant for United, was named Lisa Zanatta Henn. Her father, Peter Robert Zanatta, had landed in the first wave on Omaha Beach. "This event was probably the most important of his life," she wrote Reagan. "He also planned to go back someday." Alas, he had died of a brain tumor in 1976, and Zanatta Henn wanted to go in his place, to bring her mother and brothers to Normandy to commemorate the fortieth anniversary to honor Dad. "We would like to attend not just as tourists but as representatives of the United States," she wrote. "I don't know if there will be any special envoys to Normandy, but if there are, we would like to be a part of them. We plan to get there any way we can, but it would be nice to be part of a group of proud Americans who although they may not have been there know the anguish and pride of those who faced that day." This was all part of her cover letter; a special essay about her father was attached.

History had not remembered Private First Class Peter Robert Zanatta of the 37th Engineer Combat Battalion. He was just one of the 16 million Americans who served in uniform during World War II. Born on July 8, 1924, in Elk Creek, Idaho, Peter grew up speaking Italian at home. His family soon moved to the Potrero Hill area of San Francisco, its Tennessee Street a teeming Italian neighborhood fueled by import-export stores. Peter not only learned English but could mimic the various dialects of Italy, from Sicilian to Neapolitan. "I think he learned to pick up girls in every

Private Peter Zanatta of the Army 37th Combat Engineer Battalion. (Courtesy of Lisa Zanatta Henn)

language," Lisa joked. Although only five foot seven, Peter was muscular, and he was imbued with a deep Catholic sense of right and wrong. He was only an average student at Commerce High School, while doing carpentry work on the side. Upon graduation he was drafted into the Army. His basic training took place in Fort Pierce, Florida. "I meant to write you before but I've been very busy training down here in Florida and is this place hot," Private Zanatta wrote a friend in late 1943. "Last week I went to Palm beach where the rich people go for the winter and the place is no good. San Francisco is three times as big and the streets are smaller."

Lisa saved all of her father's wartime correspondence from both Florida and France. While his letters are often grammatically incorrect, they exude a deep humbleness and gratefulness. "A few days ago my company received a citation for outstanding work on D-Day," Zanatta wrote on October 6, 1944. "I am sorry to say that some of the boys, they had it come more than I did, are not here today to get it. But I guess their number was up. I thank the Lord that I'm still alive." Zanatta *was* one of the lucky ones who stormed Omaha Beach on June 6, 1944; he came home alive and whole, looking for a steady job and a wife. He found both. "When he came back from the war he started working as a carpenter," Lisa recalled in 2004. "He met my mom, the sweet Peggy Lou, in 1945. They were married in 1950. My brother Steve was born in 1953, I was born in 1956, and my brother Robert was born in 1958."

While some families celebrate Father's Day every June, the Zanattas always made a big deal out of the D-Day anniversary. It was their Father's Day, Independence Day, Memorial Day, and Veterans Day combined. "Dad's wartime friends would come over to our house," Lisa remembered. "And it was like a holiday: food, laughs, tears, and great conversation. The bottle of Jack Daniel's would be gone."

Cut to 1982. His daughter, Lisa, was twenty-six years old and about to get married. Over the years she had listened carefully to her father's D-Day stories, hoping that one day they would go visit the graves of his fallen friends who were buried in the American cemetery at Colleville-sur-Mer. As she picked out her wedding dress, however, she worried about the fallibility of her memory. What if over time she forgot her father's poignant stories? Didn't she owe it to her family's history to record his memories? Many children of veterans have such loving thoughts, but they're usually of the passing nature. Intuitively realizing the Sinclair Lewis maxim "The first rule of writing is to put the seat of your pants to the seat of your chair," Lisa decided not to let the moment slip away. That March, shortly after getting married, she composed a four-page "story" about her father's D-Day memories for family posterity. It was called "Someday, Lis, I'll Go Back." It was her way of honoring her deceased dad for Father's Day and the fortieth anniversary of D-Day.

SOMEDAY, LIS, I'LL GO BACK

"Someday, Lis, I'll go back and I'll see it all again. I'll see the beach, the barricades, and the graves. I'll put a flower on the graves of the guys I knew and on the grave of the unknown soldier—all the guys I fought with."

I heard my father say these words hundreds and hundreds of times for as long as I can remember. When he said them, he always looked like he was somewhere else, remembering something painful yet something he was so proud of.

My dad landed on "the beach"—First Wave, Omaha Beach, the Invasion of Normandy, June 6, 1944. The infamous D-Day. Not many people my age know or even care about this day but I always will—I can't remember when it wasn't important to me.

I know most fathers tell their kids war stories. The kids start to roll their eyes and say "oh no, not again. We've heard them all a million times." My brothers and I never said that in our house. No matter how many times we heard the stories, we never got tired of them. I tried to figure out why my dad's stories were different. The only thing I came up with is that he made you *see* it all, made you *feel* how it must have been.

My dad was eighteen years old when he went into World War II. *Eighteen*—when I was eighteen, I graduated from high school and the only heavy decisions I had to make were what college I wanted to go to or what kind of car I wanted my parents to buy me. Real life and death situations. But when my dad was eighteen he had no choices, he went and fought for his country and was proud of it. He never even thought twice about it. But those three years and the Normandy Invasion would change his life forever.

I can only remember a few of the stories he told us. There was one about a castle in Europe that had a long winding staircase. I guess my dad and his division were camping there for the night. Most of the guys were my dad's age, so being kids they slid down the banister. This always struck my brothers and me so funny—that my dad slid down some banister, in some castle in some strange city in Europe during the war. It seemed they found a moment to be kids in a situation that would turn them old before their time.

I also remember the story about how he had to lay for a long period of time on top of a dead soldier without moving as German troops plowed by. He told us of how he was afraid to breathe because the Germans might see him; of how the smell of the dead man made him so sick. We just looked at him with awe and without really comprehending it all. Not then anyway.

There were many stories—Christmas over there when the shooting stopped for a few minutes at midnight and turkey dinners fell from the sky; of giving his food to starving children so they would stop eating garbage; of being injured and then sent right back to the front; of the beauty of Paris even with the destruction of war; of the guys he knew—who lived and fought right next to him and those who died; of the songs they sung (that he taught us to sing); and of being afraid and yet going on every day—just trying to live and make it back to the glorious place called home.

But the story to end all stories was D-Day. No single incident in my dad's life ever meant more to him and I can understand why.

As I said earlier, my dad landed on Omaha Beach—on the First Wave. Even when I was small and he would tell us about D-Day, I could tell by the look in his eyes that this was different—this was the biggest thing that had ever happened in his life.

He made me feel the fear of being on that boat waiting to land. I can smell the ocean and feel the seasickness. I can see the looks on his fellow soldiers' faces, the fear, the anguish, the uncertainty of what lay ahead. And when they landed, I can feel the strength and courage of the men who took those first steps through the tide to what must have surely looked like instant death. I don't know how or why I can feel this emptiness, this fear, or this determination, but I do. Maybe it's the bond I had with my father. (I was really lucky—we never got tired of talking to each other.) All I know is that it brings tears to my eyes to think about my father as a twenty year old boy having to face that beach.

When I grew older, I read everything on D-Day I could find. As it turned out, the fact that my father lived to tell his children about it was a miracle. So many men died inside a little each time. But his explanation to me was, "You did what you had to do and you kept on going."

My dad won his share of medals. He was a good soldier and fought hard for his country. He never considered himself or what he had done as anything special. But I always did. I guess most kids put their fathers on pedestals, but I truly believe my father belonged on one. He gave up three years of his life and when he came back, everything was different. But he went on. He was just an ordinary guy, with immigrant Italian parents who never really had enough money. But he was a proud man. Proud of his heritage, proud of his country, proud that he fought in World War II and proud that he lived though D-Day.

June 6th is a special day at my family's house. When we were younger, my dad's best friend would come over, and he and my dad would just sit in our kitchen and drink and talk about old times until the early hours of the morning. They had been friends since they were eight years old and had both fought in the war.

They talked of the war of course; of their lost childhood (you can't ever be the same can you?); of the friends they had lost. Some people would say that they made too much of it or hung on to the memories too long. But how can anyone forget something like that? I never will and it all happened twelve years before I was even born.

My dad is gone now. It has been eight years. He died fighting a war against cancer. Even then the experience of D-Day was on his mind. When he was just about ready to go into surgery, I asked him how he was doing. He looked at me and said, "Lis, I feel just like I did at the Invasion of Normandy, I don't know if I'll live or die."

Maybe he made it too big a thing in his life. Maybe my family and I hang on to this part of my father's life and make it more than what it was. I've tried to make my friends understand what I feel, but they all just look at me like I'm kind of strange. Maybe if they had listened to my dad, they would feel the way I do. I guess most people my age feel that it all happened so long ago, why should they think about it.

But it was and always will be a big event. It changed everyone's lives—then and now. Everyone takes it for granted. Maybe that's what made my dad different. After he fought one of the most important battles in our nation's history, he could never take anything for granted again.

It will always affect me too. War stories, old songs, stories of the war, all of it gets to me. I know a lot of it is because my dad is gone now, and these things were so much a part of his life. But it was those events that made him the man he was—the man that came to be my father.

When I talk of Dad, I always say he landed on the First Wave at Omaha Beach. People are amazed that I even know or care about that day or event at all. But I'm just so proud of it and I always will be.

"I'm going there someday, Dad, and I'll see the beaches and the barricades and the monuments. I'll see the graves and I'll put the flowers there just like you wanted to do. I'll see the ceremonies honoring the veterans of D-Day and I'll feel all the things you made me feel through your stories and your eyes. I'll never forget what you went through, Dad, nor will I let anyone else forget—and Dad, I'll always be proud."

LISA ZANATTA HENN
MARCH 1984

The Zanatta Henn letter and Dolan speech draft eventually made their way to Dick Darman. He was particularly moved by the last five or six short paragraphs of the prose. Realizing that this was not a commonplace fan letter to file away in a forgotten cabinet—that Reagan would *want* to read it, for sure—he stuck it immediately in the Oval Office's in-box. Meanwhile, M. P. Caulfield wrote back to Zanatta Henn on May 10. He promised her Secretary of Defense Caspar Weinberger would soon send her an official invitation to Normandy—that was the good news. He then added, however, "Unfortunately, intercontinental travel and accommodations cannot be provided by the United States government."

While various White House officials appreciated Zanatta Henn's outreach, Reagan was *overwhelmed* by it. Her love of her father floored him. He approved Dolan's usage of impassioned passages from her letter in the body of his main Omaha Beach speech. Just like during his Culver City days making training films, he looked forward to telling the story of Private First Class Peter Robert Zanatta to the world. With classic Reagan flair, he would transform this Average Joe into a mythic warrior, a comprehensible symbol for millions of veteran families. Zanatta was an everyman. But to his daughter he was a singularly brave D-Day hero. "Without question Reagan understood the cross-generational aspects of the Zanatta story," Dolan recalled. "It was like a wonderful gift laid on our lap."

Once he was fully invested in the Zanatta family, the President also wanted to help Lisa out financially. "If Lisa's problem is inability to afford transportation how about an initiative to raise money for some people like this?" he scrawled on White House stationery in a note to Darman. Coming from the boss, that line was taken as more of a direct order than a question open to debate. Still, a quick investigation was undertaken, which confirmed that Private

First Class Zanatta had really fought at D-Day. He was known, in fact, as an excellent sharpshooter with a gung ho spirit. He had passed the White House–Pentagon litmus test. Money was now found so Lisa and her family could go to France, where they were to be given VIP treatment in Normandy under direct orders from President Reagan.

As previously noted, the Reagan administration was keenly aware that the Normandy speeches were to be delivered in the heat of a presidential campaign season. This fact continued to obsess Mike Deaver. While the Democratic candidates Walter Mondale and Gary Hart were battling each other in the California and New Jersey primaries, Reagan would be beamed into America's living rooms twice that day, honoring U.S. Rangers and Private First Class Zanatta and orating about freedom from tyranny and love of democracy. As stagecrafting goes, Reagan's pitch-perfect timing—i.e., the undisputable "round year" fact that it *was* the fortieth anniversary of D-Day—couldn't have been better scripted in Hollywood.

The political expediency of the entire Reagan-Normandy commemoration didn't escape the Soviet Union. Back in December, *Time* had chosen Reagan and Soviet general secretary Yuri Andropov as their "Men of the Year." But Andropov had died on February 9, succeeded by Konstantin Chernenko, and any *rapprochement* between the two superpowers was temporarily on hold. With daily regularity throughout the first week of June, *Pravda*, the official news organ of the Kremlin, launched a counteroffensive berating Reagan's planned European trip as an "American show" designed for U.S. domestic political consumption. These denunciations came on the heels of the Soviet decision to boycott the Olympics to be held that summer in Los Angeles and an announcement that a new Soviet nuclear missile would soon be deployed. Defense Minister Dmitri F. Ustinov went so far as to insist that the "historical truth" was that the Soviet-German front is what determined the Allied victory in the Second World War—not Normandy.

But while the Soviet Union complained, the countries of Western Europe, for economic reasons, embraced the fortieth anniversary as an opportunity for increased tourism. European Union travel agents, eager for U.S. dollars, organized $100-a-day, all-inclusive "liberation tours" where returning American veterans could bus it from Normandy to Berlin while listening to Benny Goodman, Glenn Miller, and the Andrews Sisters on the sound system. Besides the commercial boon, Western Europe itself was genuinely happy to celebrate the pivotal military event that led to the demise of Hitler. The big winner in tourist dollars was, of course, Normandy. Every hotel, pension, and summer home was rented out. American, British, and French flags were hung from every bistro and storefront along the former Atlantic Wall. The famed Norman hedgerows dividing pastureland and crops were given a pruning. There was even a report in the *New York Times* of a Norman entrepreneur selling World War II metal scraps for fifteen dollars a pop.

Everything went according to plan on the first leg of Reagan's journey across the Atlantic on *Air Force One*. In Ireland Reagan effectively was able to showcase his ancestral roots without its seeming too much of a political stunt. Then on June 5, President Reagan, accompanied by his wife

Nancy, lunched at Buckingham Palace with Queen Elizabeth II and Prince Philip. While Reagan was certainly not an Anglophile in speech or dress, he truly championed the "special relationship" between Great Britain and the United States. The friendship of Roosevelt and Churchill during the Second World War, he believed, was the crucial component in the Allied victory over fascism. As a consequence of the war, 30 million people had been killed; if not for Anglo-American unity that toll would have been much higher. With that historical precedent in mind, Reagan worked hard to cultivate a similar rapport with Prime Minister Margaret Thatcher, supporting her, for example, in the first half of 1982 as she grappled with the politically difficult Falklands War with Argentina. The two leaders were, of course, famously different in temperament. While he was outgoing, charming, and big-picture-oriented, she was cunning, coiled, and rather colorless. But they both shared the same conservative ideology. More than any other world leader Reagan encountered, it was Thatcher, the so-called Iron Lady, with whom he felt most at ease. She felt the same way. In her memoir, *The Downing Street Years,* Thatcher wrote that Reagan "did not suffer from the dismal plague or doubts which has assailed so many politicians in our times and which has rendered them incapable of clear decisions."

After lunching at Buckingham Palace on the fifth, the Reagans, who were lodging at Winfield House, the London residence of the U.S. ambassador, dined at 10 Downing Street that evening. Over roast lamb, Reagan, with great fanfare, regaled the officials assembled, including Thatcher, with stories about the heroism of the U.S. Army Rangers'

2nd Battalion *and* the amazing letter he had received from Lisa Zanatta Henn of California about her father. After a wave of tedious meetings in Dublin and London, Reagan, realizing the impact his upcoming commemorative speeches would have in America (thanks to television), was chomping at the bit to take to the microphones in Normandy. He was, after all, a storyteller, and Peggy Noonan and Tony Dolan had written him some great lines. But what animated Reagan the most was the fact that Herman Stein, formerly of F Company, 2nd Ranger Battalion, had that very afternoon climbed Pointe du Hoc. "Reagan was over the moon about my climbing to the top of Pointe du Hoc," Stein recalled. "I think he wished he could have done it with me."

Reagan had taken an immediate shine to Herman Stein. Rejected by the paratroopers because of poor eyesight, Stein volunteered for the 2nd Ranger Battalion. Blessed with an irascible sense of humor and fearless disposition, Stein made friends easily. Occasionally he got a little too cute. In one incident his commanding officer, Lieutenant Jacob Hill, caught him goofing off and made him do a humiliating version of the hokey-pokey dance in public. Just before shipping out of Fort Dix for Europe, he married his high school sweetheart, Lena, and wrote her constantly from Great Britain. History remembers Stein as one of the men who climbed Pointe du Hoc under intense enemy fire. Unfortunately, while working his way up a rope ladder, he was forced to witness the death of his buddy Jack Richards, a Pittsburgh Steeler football player, from a barrage of Nazi bullets. Later in the war, at Hill 400, Stein himself was wounded and received the Distinguished Service Cross.

In many ways, Stein's service was not dissimilar to that of the other Battle of Pointe du Hoc survivors. But, refusing either to stay in Florida for the fortieth anniversary or to sit on the sidelines, Stein, at sixty-three, decided to climb Pointe du Hoc once again. About a dozen Rangers serving with Special Forces in West Germany were reenacting the rope-ladder climb of Rudder's Rangers, and he wanted to be an active participant in the commemorative jubilee. Some of his buddies tried to dissuade him from rope climbing again. "Sixty-five-year-old Rangers shouldn't try to compete against the Green Berets," Len Lomell jokingly chastised him. "We're getting too old for that nonsense now."

A phalanx of media stood waiting on the Pointe du Hoc cliffs when Stein, now a roofer, joined the young Rangers on the beach to re-create the historic ascent on June 5. Although tired and out of breath, Stein reached the top, giving a bear hug to his old commanding officer, Captain Otto "Big Stoop" Masny of F Company. The embrace was greeted by a chorus of wild, raucous cheers. When asked how it felt to climb up Pointe du Hoc with a new generation of Special Forces, Stein had some mischievous fun. "All these younger guys will be all right if they just stick with it," he said. "They hug the cliff too much."

Reagan's mode of transportation from London to Normandy was the Marine One helicopter. He would not be spending the night in France. By sunset he would recross the English Channel to sleep in London—after a full day of touring Norman battle sites. Although the Normandy

American Cemetery and Memorial, run by Superintendent Phil Rivers, was technically U.S. soil, the official host of the day's events was French president Mitterrand. The Reagans, after all, were visitors to his country. Upon Deaver's insistence, however, the White House dictated where, when, and how long the Reagans would stay at any particular site. Although from the French vantage point the 2:20 P.M. arrival at Omaha Beach was the centerpiece of the daylong ceremonies, the Reagan team had, after much debate, decided Pointe du Hoc was where the oratorical fireworks would begin. Anxiously, Deaver made sure that all the key White House correspondents—men like ABC's Sam Donaldson and CBS's Bill Plante—understood that Pointe du Hoc was an important presidential moment not to be missed. On cue, conservative radio commentator Rush Limbaugh, among others, began talking up Reagan's Normandy visit to tens of millions of his fans.

Upon arriving at the Pointe du Hoc landing zone at 1:02 P.M., the Reagans disembarked from Marine One. Their schedule permitted them only one hour at the site. In the Channel, directly off Utah Beach, were U.S. destroyers, whose guns boomed throughout the day. Military helicopters hovered high above the crowds. Grasping Nancy's hand, as he always did, the President saluted men in uniform wherever he went. "Lord, you look at those cliffs," Nancy Reagan later told a *Washington Post* reporter, "and you wonder how in the world anybody made it." What truly surprised the Reagans as they wandered around was the fact that the site hadn't been flowered over like Gettysburg, Bull Run, or Antietam. The impact of both Nazi entrenchment and

President Reagan delivers his Boys of Pointe du Hoc address as U.S. D-Day veterans listen. (Courtesy of the Ronald Reagan Presidential Library)

Allied bombings remained vividly self-evident at Pointe du Hoc. A couple of grappling hooks, left from the cliff climb, were still there, not yet poached for souvenirs by thieves. The Reagans, with Len Lomell as their guide, were able to walk into a German gun emplacement bunker and stare out at the English Channel, imagining what the Nazi soldiers must have thought when they wiped the sleepers from their eyes and saw, like a terrifying mirage, the massive Operation Overlord armada of over 6,000 vessels. Large craters from Allied bombs pockmarked the entire area. Even four decades later, these gaping holes seemed desolate and menacing. They spoke of death. Like any good tourist inspecting a historic site, the Reagans were enlightened by their brief walk around. It helped clarify in their minds, in a way briefings couldn't, the magnitude of the D-Day drama. Their spirits, in fact, had been energized by the debris they saw all around them. It had happened in *their* lifetime. This is where *their* generation made a stand—and won. To Nancy Reagan, the Pointe had a very different feel from Omaha Beach, which she'd visited two years earlier. "What an incredible spot," she said. "Really, words can't describe how moving a place it is."

The 2nd Ranger Battalion veterans assembled at Pointe du Hoc that afternoon came from all over America. There was Thomas Ryan, who was a policeman in Chicago, and Thomas Rugiero, captain of the fire department in Plymouth, Massachusetts. Ralph Goranson was head of a sales company and Harvey Koehning was an electrical worker on oil wells. Some of the Rangers President Reagan would be addressing had taken advantage of the GI Bill. Frank South, for example, was a professor of physiology at the University of Delaware because of the bill. A man Reagan had heard quite a lot about, William Petty, was running a camp for underprivileged children in upstate New York but nevertheless made the transatlantic journey. Colonel Rudder's widow and daughter were at the Pointe, honored to commemorate the fortieth anniversary of the attack with the President of the United States.

Among the many other Ranger families who made the pilgrimage to the Pointe in 1984, each of which has a unique, moving story, the Wintzes of Nebraska may serve as an exemplar. Kathie Wintz Abts brought her nine children to collectively say the Lord's Prayer in memory of U.S. Army

Ranger Richard Wintz. A second lieutenant who had climbed the treacherous cliffs, Wintz always talked of bringing his wife and children to the knife-shaped promontory at Pointe du Hoc but never got around to it. A member of what author Robert Putnam calls "the long civic generation," Wintz eventually succumbed to cancer in 1981, surrounded by his family. "Dick had never talked much about his experiences, but during the last days of his life, his family convinced him to tell his story, and they recorded it," Joan Burney of the *Omaha World-Herald* reported in 1994. "They were overwhelmed. Kathie hadn't planned to go to France this year. But when the anniversary of D-Day approached, and stories of it dominated the newspapers and the broadcast media, she said 'I was just a basket case. One of the problems was I realized how naïve I was.' " Like so many children of World War II veterans, she had been sheltered by her dad, who didn't want his children to know the horrors he had seen at Normandy.

Usually speeches of any kind are forgettable. Ken Kesey, in his novel *Sailor Song*, wrote well about dull oratory where the high-profile speaker "ascended to the podium" and "spent their precious ten minutes, fumbling out the words like coins into the slot machine, hopeless and resigned." This was not the case with Reagan on this particular morning. With all those graying Rangers in front of him—not to mention D-Day families who had lost somebody dear to them forty years earlier—and a finely written Noonan speech in his pocket and on the teleprompter, he strode to the podium like a man with a mission. There was nothing boring, hokey, or mundane about his demeanor. When he saluted the flag it was done with such conviction that it made you want to stand up straight yourself, to embrace the fact that you too were part of the great American pageant. He was the American statesman about to remind the American people—with the English Channel and the Pointe du Hoc Ranger Monument at his back—what true patriotism was all about.

The entire Rangers-climbing-the-cliff story, in fact, served Reagan's worldview as a metaphor for life. Like Job, you start your ascent up the dangerous mountainside with great fortitude. But you never know *what* will knock you down, or *when* it will cripple your ascent. Life was precious. The important thing was stoically trying, one foot at a time, with God as your guide, to succeed, always heading upward to the sky. Determination and faith were what mattered. Complaints never accomplished a thing. When you fell, you picked yourself back up and tried again. With these thoughts in mind, and due to a combination of luck and design, the stage was set at Pointe du Hoc for Reagan to deliver the most remembered speech of his first term. The words Noonan had written for him that afternoon were a distillation of his anti-Communist thinking of almost four decades.

Looking the part of a world statesman, Reagan, dressed in a pin-striped suit, cleared his throat, looked directly at the wife of Colonel Rudder, and began. "At dawn, on the morning of the sixth of June, 1944, 225 Rangers jumped off the British landing craft and ran to the bottom of these cliffs," he intoned, making direct eye contact with the returning Ranger veterans. "Their mission was one of the most difficult and daring of the invasion: to climb these sheer and desolate cliffs and take out the enemy guns. The Allies had been told that some of the mightiest of these guns were here and

President Reagan shakes the hands of 2nd Ranger Battalion veterans on June 6, 1984. (Courtesy of the Ronald Reagan Presidential Library)

they would be trained on the beaches to stop the Allied advance.

"The Rangers looked up and saw the enemy soldiers—the edge of the cliffs shooting down at them with machine guns and throwing grenades. And the American Rangers began to climb. They shot rope ladders over the face of these cliffs and began to pull themselves up. When one Ranger fell, another would take his place. When one rope was cut, a Ranger would grab another and begin his climb again. They climbed, shot back, and held their footing. Soon, one by one, the Rangers pulled themselves over the top, and in seizing the firm land at the top of these cliffs, they began to seize back the continent of Europe."

Everybody in attendance was overwhelmed by Reagan's speech. Famed CBS newsman Walter Cronkite, by no means a pro-Reagan reporter, was visibly shaken by the oration. Michael Deaver, who accompanied the President to Normandy, deemed it "a home run"; a decade later he called it, along with the "Challenger Disaster" eulogy, "the best speech of his presidency." White House Chief of Staff James Baker noted that his boss that day was pitch-perfect, as if, for a few minutes, he actually *personified* the World War II generation. "I remember sitting in the audience, shaking my head, thinking, Boy oh boy, this is a dynamite moment," Baker recalled in a 2005 interview. "With Reagan what you saw is what you got. And the tears in his eyes that afternoon, believe me, they were real."

As historian Garry Wills noted in *Reagan's America,* the Pointe du Hoc speech was an example of the fortieth president offering up "the past as present." As the TV cameras flashed to the sixty-two Rangers in attendance, tears filling their eyes, it was, as Baker maintained, impossible not to be moved. These "boys" Reagan was evoking weren't just men now, they were grandparents (many had brought their grandchildren along). The power of Reagan's oration was that he spoke directly to these Rangers; in addition, he was unafraid to make eye contact with them. The message was clear: These men fought for freedom against Nazism, so don't we now have an obligation to fight against Soviet-style communism?

After his speech Reagan and his wife went and hugged all the Rangers. They then headed for Omaha Beach. There had been some last-minute debate over how much of the focus of the Omaha Beach speech should be on Lisa Zanatta Henn. On May 28 Robert Kimmitt of the NSC sent Darman a revised version of the Dolan text. "The attached draft, written by State and NSC, refocuses the speechwriter's draft—which concentrated heavily on one personal experience—toward a broad tribute to the sacrifices of the American and Allied soldiers," Kimmitt said in his cover note. "It also draws attention to the role of the French Resistance which is important given the fact that President Mitterrand will attend the ceremonies with the President."

Even as late as June 4—the day Zanatta Henn and her family arrived at the Concorde La Fayette Hotel in Paris—a minor squabble raged in the administration over what Deaver called "the desired effect" of the Omaha Beach speech. NSC and State wanted a foreign policy address aimed at enhancing the Atlantic Alliance; Deaver, Darman, and Dolan were looking for an emotional moment, one that could play well back in the United States. "They were all trying to figure out what role I was going to play," Zanatta Henn recalled. "I didn't care. I went out to Macy's and bought a new oatmeal-colored suit. I knew I was going to meet President Reagan and that was all that mattered."

On June 6, while Reagan was speaking at nearby Pointe du Hoc, the Zanatta clan arrived at a "holding tank" at the American cemetery at Omaha Beach. Lisa nervously paced about, knowing she would be meeting the President within the hour. She was almost speechless when he appeared. Without hesitation, Reagan strode right over to her as if they had known each other for years. Clutching Lisa's hand, he hugged her. "I'm so proud of you," he said. "I can only hope my kids love me as much as you loved your dad." As Secret Service milled about, the Zanattas were told they would be seated in the front row, so Reagan could have them in view. "When the President speaks, all cameras will be on *you,*" a White House aide warned Lisa. "So don't rub your face or pick your nose."

Next, President and Mrs. Reagan paid homage at the graves of fallen D-Day heroes at the Normandy American Cemetery. Every so often the Reagans would pause, and occasionally kneel, at a soldier's grave in the 172.5-acre cemetery, as if imagining the pain his death had caused a family back in Oklahoma, New Mexico, or Rhode Island. They were particularly moved by the Garden of the Missing, a semicircular wall that contained stone tablets with the names of the 1,557 "missing" who died on—or shortly after—D-Day. President Reagan particularly was taken by the inscribed words: TO THESE WE OWE THE HIGH RESOLVE THAT THE CAUSE FOR WHICH THEY DIED SHALL LIVE.

One grave at the cemetery that the Reagans sought out was that of Brigadier General Theodore Roosevelt Jr., the oldest man in the invasion. Nancy placed a bouquet of carnations and irises on the grave. President Reagan, of course, worshipped the general's father, former president Theodore Roosevelt. But he was just as enamored with the son. Fifty-six, and with a heart condition, Roosevelt Jr. nevertheless requested to go in on the first wave on D-Day. He firmly believed his presence would boost morale. In an amazing display of courage, Roosevelt Jr. marched up and down Utah Beach while it was under fire, barking out orders with a fearlessness bordering on insanity. "General Roosevelt was there, walking up and down the beach with his cane," Sergeant Richard Cassidy of Company C later recalled. "I called out, 'Go knock that bastard down, he's going to get killed.'" But Roosevelt kept parading about, inspiring the men. For his actions on D-Day, he received the Medal of Honor. Five weeks later, on July 12, he died of a heart attack. This was the kind of gutsy individual Reagan loved to evoke.

A proud, no-nonsense Reagan began his Omaha Beach speech with some elegant references to America's 9,386 graves at the Colleville-sur-Mer cemetery and then quoted General Omar Bradley, who said: "Every man who set foot on Omaha Beach that day was a hero." Reagan was standing, after all, at the water's edge, which on June 6, 1944, had turned murky red from spilled American blood. That was his lead-in for evoking the memory of Private First Class Peter Robert Zanatta. And sure enough, when his name was mentioned, the cameras, all on cue, panned in Lisa's direction. Tears were already welling up in her eyes. "Someday . . . I'll go back and I'll see it again," Reagan intoned. "I'll see the beach, the barricades, and the graves." Reagan went on to quote from Lisa's written tribute to her father, to great effect. Emotion growing in his voice, he said, "The anniversary of D-Day was always special for her family. And like all the families of those who went to war, she describes how she came to realize her own father's survival was a miracle: 'So many men died. I know that my father watched many of his friends be killed. I know that he must have died inside a little each time. But his explanation to me was, "You did what you had to do, and you kept on going."'"

Watching the Omaha Beach speech on BBC television

President Reagan, with tears in his eyes, speaking at the Omaha Beach ceremony on June 6, 1984. (Courtesy of the Ronald Reagan Presidential Library)

from his London hotel was Tony Dolan. When the camera cut to Lisa breaking up, he knew he had a winner. He had been feuding terribly with Deaver over the substance of the Dublin speech, and now the success of Reagan's remarks at Omaha Beach had given him a one-up on Mike. "You watch the tape of it and you realize it was Reagan in control even as he's struggling to get through it," Dolan later noted. "He even had Mitterrand, the original stone face, getting emotional. Something about Reagan had made his military aide bring [the letter] to him—and, of course, he responded to it. And then it was forgotten until something about Reagan made one of his speechwriters recognize the pure Reagan aspects of the letter—the Reagan gold in it. All these synergies were responsible for that draft and also speak to the culture Reagan established, a culture at the heart of which was his own very well-defined philosophy and outlook and persona."

As NSC and State had requested, Reagan did talk about the French Resistance, but it was Lisa Zanatta Henn—the daughter who loved her father so much—who captured the day's spirit. "I was just totally overwhelmed," Lisa recalled in 2004 from her home in Pleasanton, California. "Reagan gave my dad immortality. He was just an unknown private doing his duty at Omaha Beach, now his name is part of history. The whole fortieth-anniversary ceremony left me utterly speechless." Of course, by the time the Henns returned to California, the media hullabaloo had subsided, and Lisa went back to her job as a flight attendant. To her surprise and delight, however, President Reagan did not forget her. They would stay in touch for years to come. "I think he has a crush on you," Lisa's husband, Thomas, joked. "He really does wish you were his daughter."

During that same week in June 1984, Jimmy Carter's former vice president, Walter Mondale, Reagan's Democratic opponent, was campaigning around America with an FDR-JFK-LBJ liberal agenda: civil rights, Social Security, environmental protection, workers' compensation, the minimum wage, and so on. Reagan, in contrast, from Normandy, latched onto religious faith and what it meant to be an American military hero. With resolute firmness, and a good-natured shrug, Reagan fed into the growing public embrace of a World War II GI as being the exemplar of national excellence. As a cultural conservative, Reagan believed that the best way to roll back Woodstock nation was to trump it with Normandy nation. Reagan's motto, in fact, could have been,

"When in doubt, raise (or wave) the flag." For sheer oratorical elegance, the Boys of Pointe du Hoc speech—delivered, in part, to goad the Soviet Union during the Cold War—would be one of the most inspirational presidential speeches ever delivered. And the story of Private First Class Peter Robert Zanatta would get recycled in other high-profile Reagan speeches to come.

Ronald and Nancy Reagan place flowers on the grave of Brigadier General Theodore Roosevelt Jr. (Courtesy of the Ronald Reagan Presidential Library)

7

AFTER THE SPEECHES

In keeping with his amiable nature, President Ronald Reagan, following the Normandy ceremonies, corresponded with the Zanatta family. Handwritten letters were exchanged for nearly two years. "How will I ever find the words to thank you for what was probably the greatest day of my life?" Lisa Zanatta Henn wrote the President a few weeks after the Omaha Beach ceremony. "I still can't believe it all happened. Every time I think about it, I get chills. It meant so much to me to hear you say my words. My father was truly a wonderful man. I wish he could have been there to meet you, too. He loved this country with all his heart and he made us understand how blessed we were to be born in this country. I know he was with us in spirit. I will never forget any of this for as long as I live."

A cynic could argue that Reagan's Normandy speeches were entirely political, written and delivered with an election-year audience in mind. By appealing to what historian Stephen E. Ambrose among others popularized as the "we" generation, Reagan was bound to make inroads into Sun Belt states heavily populated with senior citizens, like Texas, Florida, and Arizona. There is truth to this thinking; it's not, however, the full story. As a member of the Greatest Generation, Reagan truly believed his Normandy addresses to the bottom of his heart. And listeners intuited this authenticity of purpose. His diction rang true to the ears of ordinary Americans. The *New York Times* columnist Frank Rich, hardly a Reagan enthusiast, noted shortly after the Great Communicator's death that the Boys of Pointe du Hoc speech was a "blockbuster elegy," which would prove impossible for either Bill Clinton or George W. Bush to emulate (at the fiftieth or sixtieth anniversary, respectively).

Reagan's Normandy elocutions, coupled with heavy media coverage of D-Day's fortieth anniversary, also triggered a wave of new appreciation for veterans of the so-called Good War from baby boomers. Suddenly, American community colleges, ROTC classes, lecture forums, and public libraries were inviting World War II vets to tell their own versions of what happened at Normandy, Anzio, or a hundred other dangerous places. Thunderous applause from the *Easy Rider* generation accompanied almost every recorded appearance. The History Channel started doing multipart series about various battalions and companies. These World War II veterans were sustainable heroes in a transitory *People* magazine culture that showcased sports stars and Hollywood actors as celebrities of the minute. Every American community, it turned out, had these decorated veterans in their midst. And even the grandchildren and great-grandchildren of World War II veterans—as captured in the moving last scene in *Saving Private Ryan*—were being brought into the honoring fold. "My children and grandchildren will know and feel what this day meant to us," Lisa continued in her letter to Reagan. "And as I promised, no one will ever forget the pride we all felt for all our men."

Another direct result of Reagan's evoking the memory of Private Zanatta in such a public way was that Lisa received letters from men who had served with her dad. Only five days after the President's Omaha Beach speech, Elbert E. Scudero of Antioch, California, for example, wrote Lisa a warm, loving letter. It's indicative of the kind of cross-generational enthusiasm Reagan's New Patriotism was triggering across America. By opening up the window of World War II, and saluting veterans, Reagan had started a new dialogue about a neglected war, between young and old.

Dear Mrs. Henn:

I debated some time before deciding to write but thought you might want to hear from me, so here goes.

While reading some of the D-Day articles in the June 7 edition of the Oakland Tribune *I came across the attached copy of one of the articles that related to you, your father, and the rest of your family.*

This particular article was much more than just of passing interest to me. It was the first of many I'd read over the years that specifically mentioned the 37th Engineer Combat Battalion and one of the members of that Battalion who was involved in the initial assault landing on Omaha Beach on the morning of June 6, 1944. It also brought back memories that I hadn't experienced in decades.

You see, Peter Robert Zanatta, Serial #39130513, was a member of my Company, B Company, 37th Engineer Combat Battalion. I was the Company Commander at the time of the landing, and your father was either in my boat or one commanded by Lt. Charles Peckham, both of which contained the initial groups from the assault company of engineers from the 37th to land with the Infantry and assist in securing Red Beach on Omaha. I'm sorry I can't tell you for sure which boat your father was in, but after forty years, unfortunately, some of the details of the landing become a little hazy. I do vividly recall that we landed attached to the 2nd Battalion 16th Infantry, 1st Division which were the initial assault troops to land on Reasy Red Beach. I noted that we were to assist in securing the beach, except, as I'm sure your father told you, the Germans would not cooperate without a firefight that lasted for hours, during which time we all lost some good friends and excellent men.

I was sorry to read that Peter is no longer with us and hope that this letter does not bring up any

> *past emotional memories. To lose one's parent is always a terrible blow, particularly in a close knit family, and my sympathy goes out to you and the rest of your family on the loss of your father. It's a shame that a group of good men such as we had in B Company had to drift so far apart that, with few exceptions, each one's whereabouts is unknown. This appears to be life, however, so it has to be accepted, liked or not.*
>
> *Thanks again, Mrs. Henn, for the opportunity to read in the article excerpts from your letter to your mother and President Reagan. I know it came from the heart and a love of your father. Also, thank you for fulfilling your father's wish that some day he might return to Omaha Beach to relive in part that day he could never really forget.*
>
> SINCERELY,
> ELBERT E. SCUDERO

Since becoming president, Reagan worked overtime in his crusade to conjure up a nationwide sense of patriotism based around increased defense spending. But up until June 6, 1984, his hard-line, hawkish positions weren't fully resonating with the American public—a dangerous proposition in an election year. For example, on February 9, a poll showed that only 38 percent of Americans approved of his foreign policy; 49 percent disapproved. That March, a high-ranking CIA agent was kidnapped by terrorists in Lebanon. While in a secret meeting weeks later, Secretary of State George Shultz was scolded by Nicaraguan president Daniel Ortega;

Nicaragua's internal affairs, Ortega maintained, were *not* the business of the United States. If Reagan had a foreign policy success to brag about, it was the Grenada invasion, a mere blip in the annals of U.S. military affairs.

In writing President Reagan later that summer, Lisa Zanatta Henn articulated what many felt about the new embracing of America's past military glory. "Mr. President, there seems to be a patriotic feeling sweeping our nation," she wrote. "I certainly felt it in D-Day and now with our wonderful Olympic triumph, I feel it even more. You have a great deal to do with this rekindling of love for this country. It shows in all you do. Thank you for making my father's dream come true. You saw in my words what I hoped you would see. Your speech was beautiful. For that moment in time I felt that we all had known each other for years. It says a lot for our country when a girl from San Francisco can feel that the President of the United States is a close friend."

Lisa had ended her letter offering her own family's diehard support for Reagan's reelection effort against Walter Mondale. They would go door-to-door for the Gipper in Marin County, handing out bumper stickers and buttons. "Our thoughts and prayers will be with you through what must be a very tiring time," she wrote. "And after the *next four years,* if you do retire to California, please know that you always have friends to call on. We all feel so close to you. It was the greatest of honors to meet you." It was signed "All My Love—Lisa Zanatta Henn."

Lisa's mother, Peggy, also wrote President Reagan a long letter of gratitude. He replied: "Where do I find the words to thank you for your beautiful letter and for the photo which

Lisa Zanatta Henn greets President Reagan in Normandy for the first time. They developed an epistolary relationship. (Courtesy of Lisa Zanatta Henn)

I'm very proud to have. Meeting all of you and sharing that day with you made it a very special day for us, one we shall always remember with great warmth."

Reagan, of course, easily defeated Mondale in November 1984 to win a second term. Senior citizens made up a major component of his constituency. But for a few months, he almost lost credibility with members of the "we" generation over a controversial trip to West Germany. Always worried about keeping Bonn vigilant in the Cold War showdown with the Soviet Union, Reagan accepted an invitation from Chancellor Helmut Kohl to visit Bitburg cemetery during his upcoming May 1985 visit to Europe. No sooner was this announced than Reagan found himself being lambasted by the media and various Jewish watchdog organizations. Bitburg, after all, contained graves of Nazi soldiers of the

Waffen SS. Was he really going to place a wreath on the grave of a Nazi? Was pleasing Chancellor Kohl worth engaging in such a demeaning venture? Columnist Richard Cohen of the *Washington Post* summed up the reaction of most people to the ill-advised Bitburg visit: "It is an obscenity even to suggest that a president of the United States honor those men. You do not honor the present generation of Germans by ignoring the crimes of the previous one. . . . World War II had a moral component that cannot be overlooked. It is not yet some distant contest in which we can trot out clichés about the 'honored dead'—as if the honor is in dying and not the reason for it. The dead at Bitburg died defending Nazism. To honor them does not dishonor their victims—nothing can do that. It merely dishonors."

By merely announcing he was going to Bitburg, President Reagan unleashed a firestorm of protest, a media backlash that could have potentially tarnished his newfound credentials as spokesman for the World War II veterans. A public relations debacle was at hand. Deaver, the whiz choreographer of Reagan trips to Ireland, France, South Korea, and other countries, had made a misstep. Due to Reagan's Bitburg embrace, flowers were already being placed on the graves of SS soldiers and some of the German far right were deeming these Nazis victims of Hitler's mistaken policies. Reagan thought about canceling the Bitburg visit, but on May 5, ignoring his critics, he visited both the cemetery and the Bergen-Belsen concentration camp. His hope was to celebrate reconciliation between the United States and West Germany.

Of all the missives that Reagan sent Lisa and her family,

his letter of May 22, 1985, stands out. Worried about how World War II veterans—and their families—would view his Bitburg visit, he tried to explain his rationale to his new California friends. "I've heard from a number of veterans who, like your father, had stories to tell about moments in the war when deeds of human kindness momentarily bridged the gap between enemies," Reagan wrote. "One thing that made me want to do the trip was the knowledge that in these forty years since V.E. Day the Germans have become our friends and allies and have never asked us to forgive or forget the Nazi evil. Indeed, they take their school children to the concentration camps and show them the horror of the Holocaust. They do it so such a thing can never happen again."

Reagan continued to recount Lisa's story, "Someday, Lis," to others every time World War II came up. As noted, what captivated him about the Zanatta tale was its family-oriented, cross-generational aspect. Often estranged from his own daughters, Maureen and Patti, it was as if he hoped Lisa's loving disposition toward her deceased father would rub off on them. As men grow older, into their sixties and seventies, they often hope that their daughters will be caretakers for them. It is a tradition that, of course, runs deep in millions of American families like the Zanattas.

Although some claim that Father's Day was created in West Virginia in 1908, the true progenitor of the holiday was a daughter, Sonora Smart Dodd of Spokane, Washington, who wanted to honor her dad's gifts as a child raiser. Inspired by Anna Jarvis's creation of Mother's Day, on June 5, 1909, her father's birthday, Dodd asked her minister to prepare a special celebratory service to celebrate her father,

William Smart, a farmer and Civil War veteran. His wife had died giving him their sixth child. Widowed, he managed to raise the brood by himself. With Sonora's lobbying, a date was selected, June 19, 1910, to honor her dad at church. The idea spread. The following year, the mayor of Spokane and the governor of Washington officially supported the event. It soon spread across America. By 1924 President Calvin Coolidge publicly supported a plan for a national Father's Day; it was not, however, until 1966 that President Lyndon Johnson signed a presidential proclamation mandating that the third Sunday of every June be Father's Day.

Private Zanatta's World War II service—and its honorific Father's Day–like aspect—provided Reagan with a real-life narrative text he could embrace from four different angles. First, it was the son-of-immigrants-makes-good-in-America saga personified. Call it a low-keyed, true-to-life Horatio Alger tale. Second, and most important, it was about celebrating patriotism vis-à-vis military service. Clearly Zanatta's love of America was deep and genuinely felt. Third, Zanatta's wartime service had a liberating effect on the young soldier. Always looking for the optimistic silver lining in a story, Zanatta, unlike the men buried at Colleville-sur-Mer, survived World War II without serious injury. His military service had become a positive, character-building, defining moment in his life. If Zanatta had not fought on Omaha Beach, he would have just been an insurance salesman in California. But by storming the beach, he became a man, a timeless hero to his wife and children. Finally, it does not take a psychoanalyst to point out that Reagan had felt guilty about not being a great father. (He was a great husband.) His hope, like all fathers' hopes, was that his own children would someday realize that when it came to parenting he did the best he could.

An astute interpreter of Reagan, Tony Dolan knew all of these things about his boss. And in coming years the story of Lisa Zanatta Henn was retold in a number of Reagan speeches and dinner toasts. Most famously, in 1986, when Reagan was speaking at the Statue of Liberty centennial in New York, he used the Italian immigrant tale of Private Zanatta as one of his speech's hooks. "I was still flying with United Airlines," Zanatta Henn recalled. "And I was stunned that our family's story was still part of President Reagan's appeal in our nation."

But the most moving use of Zanatta Henn came in Reagan's Farewell Address. To Lisa's astonishment, as she watched TV with her three-year-old daughter, Cassie, in Foster City, California, President Reagan's public good-bye included her D-Day story. In fact, she was a centerpiece of his Dolan-written remarks. "But now, we're about to enter the nineties, and some things have changed," Reagan said on January 11, 1989. "Younger parents aren't sure that an unambivalent appreciation of America is the right thing to teach modern children. And as for those who created the popular culture, well-grounded patriotism is no longer the style. Our spirit is back, but we haven't re-institutionalized it. We've got to do a better job of getting across that America is freedom—freedom of speech, freedom of religion, freedom of enterprise. And freedom is special and rare. It's fragile; it needs [protection]." Then he evoked the Zanatta family. "So, we've got to teach history based not on what's in fashion, but what's important—why the Pilgrims came here, who Jimmy Doolittle was, and what those thirty seconds over Tokyo meant," he said. "You know, four years ago on the fortieth anniversary of D-Day, I read a letter from a young woman writing to her late father, who'd fought on Omaha Beach. Her name was Lisa Zanatta Henn, and she said, 'We will always remember, we will never forget what the boys of Normandy did.'* Well, let's help her keep her word. If we forget what we did, we won't know who we are. I'm warning of an eradication of the American memory that could result, ultimately, in an erosion of the American spirit. Let's start with some basics: more attention to American history and a greater emphasis on civic ritual."

Nobody had warned Lisa that Reagan was going to evoke her in his Farewell Address. "It was amazing," she recalls. "I was watching Reagan speak and then he mentioned me. My daughter Cassie said, 'Mama, that man just said your name.' It was the cutest thing. I was totally shocked. I knew I had made an impression on him but *wow!*" The Zanatta Henns would never see Reagan again, but their home remained filled with photographs of him hanging on the wall, as if he were an adopted member of their nuclear family.

The Zanattas weren't the only new friends Reagan made over Normandy. Due in part to the success of the Boys of Pointe du Hoc speech, Richard Darman got Peggy Noonan invited to an Oval Office meeting. She had been at the White House four months and her request had finally come to fruition. When Noonan entered his office, Reagan was

*Lisa Zanatta Henn never wrote this line. President Reagan was summarizing her sentiment.

ensconced behind his desk, all smiles, adjusting a large hearing aid that he wore. "He twinkled at me," Noonan wrote. "I was the new one, and the only woman. He walked to me and took my hand. It is the oddest thing and even if everyone says it: It is not possible to be nervous in his presence. He acts as if he's lucky to be with you. 'Well,' he says, 'it's so wonderful to meet you, please, sit down, well, so.' " With trademark, good-natured humor, Noonan, in her memoir, vividly details what the meeting was like. She didn't even know for sure if Reagan knew who she was—he didn't. At meeting's end, however, Reagan held Noonan's hand, escorting her to the door. Ben Elliott, coming to Noonan's rescue, informed Reagan that Noonan was the one who had written the Pointe du Hoc speech. Then he not only twinkled, he lit up like a Christmas tree. "That was wonderful," Reagan cooed, "it was like Flanders Fields.* I read it upstairs, and when I read something I like to look up at the corner to see the name, and I saw Noonan. I meant to call you and never . . ."

Noonan left the White House in 1986, after only two years of speechwriting service, in part over personality disagreements with various members of the White House staff. But not before she wrote the President's moving tribute to the seven astronauts—Francis "Dick" Scobee, Michael J. Smith, Ronald McNair, Ellison Onizuka, Gregory Jarvis, Judith Resnik, and Christa Corrigan McAuliffe—killed in the *Challenger* shuttle disaster. "The crew of space shuttle *Challenger* honored us by the manner in which they lived their

*Reagan was referring to "In Flanders Field," a poem written by Canadian surgeon John McCrae about the calm of a World War I battlefield after fighting is done.

lives," Noonan wrote for Reagan. "We will never forget them, nor the last time we saw them—this morning as they prepared for their journey and waved good-bye, and slipped the surly bonds of earth to touch the face of God." With that speech, she assured herself a top-tier place in the White House speechwriter Hall of Fame. She went on to write George H. W. Bush's Thousand Points of Light speech, and continues to write for various publications such as the *Wall Street Journal, Forbes,* and *Time*. She got married in 1985 to economist Richard Rahn and had one son before getting divorced in 1989. She now lives in Brooklyn.

Besides Lisa Zanatta Henn and Peggy Noonan, Reagan made yet another new friend in the wake of his Boys of Pointe du Hoc speech. One afternoon author Roger Kahn received an unexpected telephone call from Ken Duberstein at the White House. The President loved *The Boys of Summer* and hoped Kahn, who lived in Westchester County, New York, could come visit him in the White House. An old Cubs baseball announcer, Reagan wanted Kahn to watch a 1952 Warner Brothers love story he had starred in with Doris Day, *The Winning Team*. It was a biopic of the early-twentieth-century pitcher Grover Cleveland Alexander, who was afflicted with epilepsy and alcoholism but was able to transcend his hardships and triumph in the 1926 World Series. It was a tough time for Kahn, who had just lost a child to heroin, so the Reagan offer to visit the White House lifted his sagging spirits. "I was actually quite nervous," Kahn recalled. "What was I going to say to him?"

The theme of Kahn's book was that while the 1952 Dodgers didn't defeat the Yankees in the World Series, the

President Reagan greets Roger Kahn, the author of *The Boys of Summer,* along with baseball great Joe Garagiola (behind the President). (Courtesy of Roger Kahn)

way they handled defeat made them, in the long run, winners. Reagan had a different problem. He *was* the Yankees; he had been reelected to a second term by a margin of more than 16 million popular votes, and his 525 electoral votes were among the most ever amassed in history. So Kahn, while dressing at his hotel, decided to wear his Brooklyn Dodgers Hall of Fame pin on his lapel to 1600 Pennsylvania Avenue as an icebreaker. This would trigger, he hoped, a lively baseball conversation with the President. Arriving at the White House screening room at noon, Kahn eyed ten or twelve people waiting for the film to start. Reagan was sitting by himself. When introduced, Reagan asked him a question. "Did you know that I used to broadcast baseball?" Kahn said yes, and they were off and running. They spoke about the Boys of Pointe du Hoc speech, why Jackie Robinson was a Republican, and mainly about Reagan's favorite baseball player, Grover Cleveland Alexander. A bond was quickly forged between the two men with baseball the common denominator. "About the movie—I probably talked too long at show's end—but I failed to explain one little touch in the film: the two or three times Alex was shown stuffing one stick of gum after another into his mouth," Reagan wrote Kahn on July 26, 1988. "Chewing tobacco was common among ballplayers, but Alex thought it was a dirty habit. His answer was to put all five sticks of gum in the pack into his mouth at the start of the game. Hardly a great historical fact, but I thought you should know."

Three years after the Boys of Pointe du Hoc speech, Reagan delivered another stunning address in Europe, this time in front of the Brandenburg Gate. On June 12, 1987, in strong words aimed at Soviet leader Mikhail Gorbachev, Reagan implored the Kremlin to dismantle the Berlin Wall. In the most memorable line of the speech, Reagan declared: "General-Secretary Gorbachev, if you seek peace, if you seek prosperity for the Soviet Union and Eastern Europe, if you seek liberalization, come here to this gate. Mr. Gorbachev, open this gate. Mr. Gorbachev, tear down this wall."

The story of how Reagan came to deliver his freedom-tolling speech in Berlin is best told by former White House speechwriter Peter Robinson in his memoir, *How Ronald Reagan Changed My Life* (2003). In May 1987, Robinson, now a fellow at Stanford University's Hoover Institute, was assigned to draft an address for President Reagan's upcoming journey to Berlin, a teeming city that was celebrating its 750th anniversary. Unfortunately, Berlin—although one of the blessed centers of Europe, with its Bauhaus architecture, imposing Reichstag, Tiergarten park, and smoky cabarets—was divided by a concrete barrier and encircled in barbed wire. The Berlin Wall, erected in August 1961, was a monstrous affront to Jeffersonian-Hamiltonian democracy, human rights, common decency, and laissez-faire capitalism. It had become an ugly, drab symbol of Soviet totalitarianism gone haywire. Obviously any society that had to wall in its citizens or, as Reagan put it, "had to pen its people up like farm animals," was committing an enormous affront to the very notion of justice.

Not content to just sit at his desk to draft such an important speech, Robinson flew to Berlin, took the pulse of the city, and asked a lot of questions. It was at a dinner party hosted by Dieter Elz, a former World Bank official, however, that Robinson came up with the simple but powerful "tear down this wall" phrase. (However, Tony Dolan, debunking Robinson, recalls Reagan casually saying those words in White House meetings throughout his presidency.) When Robinson asked the party about the wall, his hostess, Ingeborg Elz, suddenly made a fist with one of her hands and slapped it into the palm of her other and said, "If this man Gorbachev is serious with his talk of *glasnost* and *perestroika*, he can prove it. He can get rid of this wall."

Robinson had the line he was looking for, the centerpiece of the thirty-minute speech Reagan would deliver. "When I sat down to write, I'd like to be able to say, I found myself so inspired that the words simply came to me," Robinson wrote. "It didn't happen that way. *Mr. Gorbachev, tear down this wall.* I couldn't even get that right. In one draft I wrote, 'Herr Gorbachev, bring down this wall,' using 'Herr' because I somehow thought that would please the President's German audience and 'bring' because it was the only verb that came to mind." Eventually Robinson got it right. When Reagan read a draft of the speech he loved it, particularly the part about the wall having to come down. He had told Tony Dolan that, more than any other phrase, that's what he wanted to say in Berlin. But the State Department and National Security Council were in an uproar. They pleaded with the President to drop the inflammatory line about the wall, which they considered antagonistic in the extreme. A flurry of telephone calls and memoranda circulated, insisting that the speech be thrown away, or at least seriously rewritten. Robinson called this negative reaction he experienced "squelchfest" in *How Ronald Reagan Changed My Life*.

America's top foreign policy experts were vehement that Reagan not deliver this so-called crude, unduly provocative speech, which would only incite friction with the Kremlin. Even on the morning that Reagan arrived in Berlin, top aides pleaded with the President not to deliver the Robinson speech. Reagan told his top advisors that he would consider their recommendation. But on the limousine ride to the Brandenburg Gate, Reagan told Ken Duberstein that he just had to deliver the powerful line about tearing down the wall. With an "aw shucks" smile, he poked Duberstein in the ribs and said, "The boys at State are going to kill me, but it's the right thing to do."

Ronald Reagan had the self-confidence to pull off a clarion call for democracy on that June afternoon, just as he had on June 6, 1984. The Normandy speeches, plus the Berlin address, now stand as the rhetorical bookends to Reagan's two terms in office. His voice on these occasions offers the best statements and sentiments of his New Patriotism. It's impossible to watch Reagan deliver these speeches in TV clips and not be moved by their sheer oratorical power. They are the enduring capstones of his long, anti-Communist crusade.

Because of President Ronald Reagan's 1984 Normandy speeches, a whole new generation of Americans learned about D-Day. (Courtesy of Lisa Zanatta Henn)

EPILOGUE

When President Ronald Reagan left Normandy on Marine One on June 6, 1984, after delivering the Pointe du Hoc and Omaha Beach speeches, he was euphoric. Like any good performer, he knew when he had done his job well. But Mike Deaver was even more jazzed up. His stagecrafting had gone beyond expectations. The fifteen-foot Ranger dagger memorial wasn't just a smart campaign-year backdrop, it proved spectacular. The Pointe du Hoc speech wasn't Reagan apocrypha, like so many of his World War II stories, but a pivotal oratorical moment in remembrance, an articulate Cold War era veneration of World War II GIs that, thanks to Peggy Noonan, encompassed both reveille and taps. Add to the mix the sad, mournful lament of Lisa Zanatta Henn, an emotion-driven Tony Dolan masterstroke, and there was reason for jubilation in Reaganland. Deaver knew, then and there—as the Reagans were flying over the English Channel back to London—that the Boys of Pointe du Hoc speech, in particular, was a seminal salvo in promoting the New Patriotism. Reagan's 1984 campaign slogan, after all, was "It's Morning Again in America." As the GOP television commercials boasted: "Life is better, America is back. And people have a sense of pride they never felt they'd feel again." The last time America "felt" that way, the commercial implied, was during the Second World War or, perhaps, back in the 1950s, when Dwight D. Eisenhower was president.

Given Reagan's penchant for historical symbolism, the icing on the cake of his Normandy day was flying over the USS *Eisenhower,* positioned in the middle of the Channel. The crew of the nuclear-powered aircraft carrier, named after America's thirty-fourth president, assembled on the deck when Marine One hovered over and spelled out the name "IKE." From his high-altitude perch, Reagan chuckled and snapped a salute. "I like Ike," he shouted over the radio to the ship's captain. "I love Ike."

Earlier that day, at Omaha Beach, Reagan had talked with the venerable CBS newsman Walter Cronkite, a D-Day scholar, about his idolatry of Eisenhower. With Cronkite serving as media middleman, the D-Day baton was symbolically passed that afternoon from the battlefield legacy of Eisenhower to the presidential oratory of Reagan. "I will forever stand for Ike's D-Day veterans," Reagan told Cronkite.

"They are my heroes, like they were his." Often overlooked by historians and pundits—who prefer comparing Reagan to Barry Goldwater because of their shared conservative values—is how much the Gipper stylistically modeled his presidency after Eisenhower's. If Reagan had adopted the thespianism of Franklin D. Roosevelt as an emblem, he had also hijacked Eisenhower's infectious grin, gregarious demeanor, and "hands off" style of leadership as his own. (At least, that is, for public consumption.) After carefully studying Eisenhower's subdued Oval Office antics back in the mid-1950s, Reagan concluded that simplicity and kindness were sterling virtues in American politics. What biographer Peter Lyon wrote about Eisenhower—the only other two-term Republican president of the twentieth century—could just as easily have been written about Reagan: "He wanted people to like him; he was distressed when it failed to happen. His need for friendly rapport was one reason for his reluctance, so often marked by journalists, to speak ill of anyone."

Up until his death on March 28, 1969, Eisenhower was D-Day personified, the heroic Supreme Allied Commander who had uttered the immortal "Okay, let's go." He had built his political career around that landmark moment. "His place in history was fixed as night fell on the Normandy beaches on June 6, 1944," Stephen Ambrose wrote in *Eisenhower: Soldier and President*. "Hundreds of thousands, indeed millions of men and women contributed to Overlord, and 200,000 soldiers, sailors, and airmen participated directly on D-Day itself, but the operation will forever be linked to one man, Dwight Eisenhower, and rightly so. From inception to completion, it bore his personal stamp.

D-Day heroes Jack Kuhn and Len Lomell get together to commemorate the fiftieth anniversary of the Battle of Pointe du Hoc on June 6, 1994. (Courtesy of JoAnna McDonald)

He was the central figure in the preparation, the planning, the training, the deception, and the organization of the greatest air and sea armada in history. At the decisive moment, he was the commanding general who, standing alone, weighed all the factors, considered the alternatives, listened to the conflicting views of his senior subordinates, and then made the decision—and made the right one."

When Eisenhower first ran for president in 1952, his campaign was based largely on the fact that he was the mastermind of D-Day, the amiable general who, as the title of his memoir *Crusade in Europe* suggests, liberated a continent from Hitler's Fascist grip. D-Day was truly America's finest hour—the apotheosis of warfare—and Eisenhower knew it. By using the grandiose invasion as his political calling card, Eisenhower had transformed the Republican Party from the *losers* of the Roosevelt-Truman era to the landslide *winners* of the 1950s, in firm control of the executive branch. The Democratic nominee in both 1952 and 1956 was Adlai Stevenson, a well-reasoned liberal intellectual senator from Illinois, who never had a running chance when matched up against an all-American war chieftain like Eisenhower. In similar fashion, in 1984, the Democratic nominee, Walter Mondale, tagged with having been vice president during the Carter "malaise" years, couldn't compete with Reagan's "It's Morning Again in America" approach, of which the Normandy speeches were centerpieces. Riding on a "D-Day remembered" wave, Reagan, like Eisenhower, easily defeated his Democratic opponent to win a second term as president. Reagan won forty-nine states, to Mondale's one. In American politics, triumphalism, Reagan believed, always beat malaise.

Without question, Reagan's adoption of World War II, his picking up of the laurels where Eisenhower left off, was a political masterstroke. From 1984 onward the Republican Party, following Reagan's lead, became *the* political party of the Second World War. It didn't matter that George McGovern was a highly decorated World War II bomber pilot or that Jimmy Carter had the second-longest military career of all twentieth-century presidents or that John Glenn displayed more military heroism in a single flight day than Reagan had in an entire lifetime; the die had been cast. The selling of patriotism—like the kind the 2nd Ranger Battalion had exhibited at Pointe du Hoc—became the GOP's stock-in-trade. In 1988 and 1992 (George H. W. Bush) and 1996 (Bob Dole), the Republican Party nominated World War II heroes for president, winning one out of the three elections. The Democratic Party had become the voice of labor, environmentalism, education, the minimum wage, and Clintonism. The Republicans added war veterans and military families (along with big business and Christian fundamentalists) as key pillars in their conservative coalition.

So it was poetic irony, or fate, that Reagan died on June 5, 2004, at his home in Los Angeles. It was exactly sixty years from the day when Eisenhower had greenlighted the D-Day invasion, a fact all the international media—gathered for commemoration ceremonies along the Normandy coast—seized upon. Reagan's timing in death, in fact, was so right on cue that many newspersons muttered under their breath that the Nancy Reagan–Mike Deaver team had staged it to coincide with both the twentieth anniversary of his Boys of Pointe du Hoc speech and the sixtieth anniversary of D-Day. They had "pulled the plug," so to speak, to have Reagan die on June 5 in order to perpetuate his historical standing as a World War II–era icon. This unprovable Hemlock Society–like rumor was given further credence by the fact that since the attempted assassination of Reagan on March 30, 1981, with the fortieth president's consent, Nancy had micromanaged all the plans for her husband's funeral, making detailed adjustments to the various ceremonial aspects of the dreaded day with exacting care. Ronnie and Nancy were both in agreement that there should be funeral ceremonies in both California and Washington, D.C.

Not that Reagan's death came as a shock. Reagan had been suffering terribly from Alzheimer's disease from at least

1994, when he had written an open letter to the American public speaking candidly about his affliction. Today, the original handwritten explanation is on view under glass at the Reagan Library. "At the moment I feel just fine," he wrote. "I intend to live the remainder of the years God gives me on this earth doing the things I have always done. I will continue to share life's journey with my beloved Nancy and my family. I plan to enjoy the great outdoors and stay in touch with my friends and supporters. . . . Unfortunately, as Alzheimer's disease progresses, the family often bears a heavy burden. I only wish there was some way I could spare Nancy from this painful experience. When the time comes I am confident that with your help she will face it with faith and courage. . . . In closing let me thank you, the American people, for giving me the great honor of allowing me to serve as your President. When the Lord calls me home, whenever that may be, I will leave with the greatest love for this country of ours and eternal optimism for its future. I now begin the journey that will lead me into the sunset of my life. I know that for America there will always be a bright dawn ahead."

News of Reagan's death reached President George W. Bush in Paris on the eve of the sixtieth anniversary of D-Day. It was as if, even in death, Reagan had upstaged him. "Some would argue that no politician in his right mind would invite comparisons to the Great Communicator," columnist Frank Rich wrote in the *New York Times*. "In the aftermath of Reagan's death, his fans and foes alike remain agog at his performance chops." All of the major news networks—ABC, NBC, CBS, CNN, and Fox News—went into emergency mode, forced to change the programming

President Ronald Reagan died on June 5, 2004, from Alzheimer-related causes. Here he is alongside French president François Mitterrand in 1984, placing wreaths at the Omaha Beach Memorial. (Courtesy of the Ronald Reagan Presidential Library)

from "Bush at D-Day plus Sixty Years" to "Reagan Dead at Ninety-three." The preeminent sound bite, along with "Tear down this wall," was the Boys of Pointe du Hoc speech. It was as if Rudder's Rangers had become Reagan's Rangers. "A great American life has come to an end," Bush said about Reagan's demise. "America laid to rest an era of division and self-doubt and because of his leadership the world laid to rest an era of fear and tyranny." Writing in the *New York Times,* reporter Marilyn Berger accurately invoked how Reagan, during his eight-year presidency, "managed to project the optimism of Franklin D. Roosevelt" with the "faith in small-town America of Dwight D. Eisenhower." She added that Reagan was "imbued with a youthful optimism rooted in traditional virtues of a bygone era."

President Bush, at least for this day—and some would argue for his entire presidency—found himself in the shadow of Ronald Reagan, the president who, as Margaret Thatcher claimed, "won the Cold War without firing a single shot." As the television networks played Reagan's the Boys of Pointe du Hoc speech over and over, President Bush's halting Normandy address paled in comparison. (So would have one delivered by Bill Clinton.) As U.S. ships boomed their guns, and World War II veterans wept, thousands gathered at Colleville-sur-Mer cemetery to pay homage to Reagan, whose body lay in a California mortuary. Along with the men of D-Day, he was being honored on equal footing. Even NBC News anchor Tom Brokaw, the originator of "the Greatest Generation," on the air live from the cemetery, realized that it wasn't just a two-term president who died, but the "Great Communicator" who rose to personify, in his senior years, the entire World War II generation of which he had been a charter member.

Once June 7 came around, however, and the D-Day ceremonies ended, other aspects of Reagan's long life started eclipsing the boys of Pointe du Hoc. Television pundits debated the perils of Reaganomics, the Iran-contra scandal, and the President's racial insensitivity. While some critical perspective was added to the television coverage by Democrats, for the most part, however, the media lovefest for Reagan continued unabated for a week. Nancy Reagan had brilliantly planned a state funeral for her husband, the first in thirty years. All the stops were pulled out: horse-drawn caisson, flag-draped coffin in the Capitol's rotunda, and a 138-page "planning booklet" that covered everything from seating charts to floral arrangements. Since her husband's 1994 announcement of Alzheimer's, the former First Lady had become an advocate for intensified medical research about the ailment, a bold spokesperson for the over 4 million stricken with the horrible disease. "Ronnie's long journey has finally taken him to a distant place where I can no longer reach him," she said. "Because of this, I'm determined to do whatever I can to save other families from this pain."

Many historians have thrown their hands up in bafflement about what made Reagan tick. Truth be told, Reagan was not an enigma. He was a hardworking actor who created the persona of Ronald Reagan. By adopting attributes of FDR and Eisenhower and the gee-whiz, brave, good-guy shrug of the World War II GI, along with a healthy dose of mythic cowboy self-reliance, Reagan, in death, had become an American archetype. Such an amalgamation of character traits seemed hokey back in the 1950s, when the brooding existential angst of Marlon Brando and James Dean was in vogue. But by the late 1970s, when America was exhausted by the incessant turmoil of Vietnam, civil rights, and Watergate, Reagan came back on the American scene with a nostalgic vengeance. John Wayne had cornered the market on the cowboy mystique, but being the impresario of the World War II generation—a role Tom Brokaw and Tom Hanks would assume in the 1990s—was up for grabs. He seized upon the role with relish. His two 1984 Normandy speeches—Pointe du Hoc and Omaha Beach—sealed the deal for his ascendancy as the clear, logical spokesperson for the self-effacing World War II veterans.

Lisa Zanatta Henn was watching *Scarborough Country* on MSNBC from her home in California when the news of Reagan's death flashed on television. She had wanted to attend the sixtieth anniversary of D-Day commemorations in Normandy but decided to stay home because her daughter, Cassie, was graduating from Foothill High School in Pleasanton. Within moments of the announcement, MSNBC played a pretaped and skillfully edited newsmagazine profile of Lisa's relationship with Ronald Reagan. It began with a photograph of Peter Zanatta in uniform. A few seconds later it showed her tears at Omaha Beach back in 1984. Her story had become, at least on one cable network, the lead story of Reagan's life. She started crying again. She always felt safe when Reagan was alive; now there was a terrible void. "He was, for my generation, the patriarch of World War II," she later noted. "While Patton, Eisenhower, MacArthur, and all the others were bigwigs, Reagan represented the regular soldier, the GI Joes. He was their voice of remembrance." Then, catching her breath, she said a line Reagan would have truly appreciated. "He cared about our boys more than anybody else I ever knew."

That Reagan was able to convince not just Lisa Zanatta Henn but millions of World War II veterans' families of his deep empathy for them was one of his greatest political accomplishments. The story of D-Day as the pervasive metaphor for American bravery and goodness, in part because of his presidential voice, endures for the ages to ponder.

APPENDIX

Ronald Reagan's Speech at Pointe du Hoc

Commemorating the Fortieth Anniversary of the Normandy Invasion, June 6, 1984

We're here to mark that day in history when the Allied armies joined in battle to reclaim this continent to liberty. For four long years, much of Europe had been under a terrible shadow. Free nations had fallen, Jews cried out in the camps, millions cried out for liberation. Europe was enslaved, and the world prayed for its rescue. Here in Normandy the rescue began. Here the Allies stood and fought against tyranny in a giant undertaking unparalleled in human history.

We stand on a lonely, windswept point on the northern shore of France. The air is soft, but forty years ago at this moment, the air was dense with smoke and the cries of men, and the air was filled with the crack of rifle fire and the roar of cannon. At dawn, on the morning of the sixth of June, 1944, 225 Rangers jumped off the British landing craft and ran to the bottom of these cliffs. Their mission was one of the most difficult and daring of the invasion: to climb these sheer and desolate cliffs and take out the enemy guns. The Allies had been told that some of the mightiest of these guns were here and they would be trained on the beaches to stop the Allied advance.

The Rangers looked up and saw the enemy soldiers—the edge of the cliffs shooting down at them with machine guns and throwing grenades. And the American Rangers began to climb. They shot rope ladders over the face of these cliffs and began to pull themselves up. When one Ranger fell, another would take his place. When one rope was cut, a Ranger would grab another and begin his climb again. They climbed, shot back, and held their footing. Soon, one by one, the Rangers pulled themselves over the top, and in seizing the firm land at the top of these cliffs, they began to seize back the continent of Europe. Two hundred and twenty-five came here. After two days of fighting, only ninety could still bear arms.

Behind me is a memorial that symbolizes the Ranger daggers that were thrust into the top of these cliffs. And before me are the men who put them there.

These are the boys of Pointe du Hoc. These are the men who took the cliffs. These are the champions who helped free a continent. These are the heroes who helped end a war.

Gentlemen, I look at you and I think of the words of Stephen Spender's poem. You are men who in your "lives fought for life . . . and left the vivid air signed with" your honor.

I think I know what you may be thinking right now—thinking "we were just part of a bigger effort; everyone was brave that day." Well, everyone was. Do you remember the story of Bill Millin of the 51st Highlanders? Forty years ago today, British troops were pinned down near a bridge, waiting desperately for help. Suddenly, they heard the sound of bagpipes, and some thought they were dreaming. Well, they weren't. They looked up and saw Bill Millin with his bagpipes, leading the reinforcements and ignoring the smack of the bullets into the ground around him.

Lord Lovat was with him—Lord Lovat of Scotland, who calmly announced when he got to the bridge, "Sorry I'm a few minutes late," as if he'd been delayed by a traffic jam, when in truth he'd just come from the bloody fighting on Sword Beach, which he and his men had just taken.

There was the impossible valor of the Poles who threw themselves between the enemy and the rest of Europe as the invasion took hold, and the unsurpassed courage of the Canadians who had already seen the horrors of war on this coast. They knew what awaited them there, but they would not be deterred. And once they hit Juno Beach, they never looked back.

All of these men were part of a roll call of honor with names that spoke of a pride as bright as the colors they bore: the Royal Winnipeg Rifles, Poland's 24th Lancers, the Royal

Scots Fusiliers, the Screaming Eagles, the Yeomen of England's armored divisions, the forces of Free France, the Coast Guard's "Matchbox Fleet" and you, the American Rangers.

Forty summers have passed since the battle that you fought here. You were young the day you took these cliffs; some of you were hardly more than boys, with the deepest joys of life before you. Yet, you risked everything here. Why? Why did you do it? What impelled you to put aside the instinct for self-preservation and risk your lives to take these cliffs? What inspired all the men of the armies that met here? We look at you, and somehow we know the answer. It was faith and belief; it was loyalty and love.

The men of Normandy had faith that what they were doing was right, faith that they fought for all humanity, faith that a just God would grant them mercy on this beachhead or on the next. It was the deep knowledge—and pray God we have not lost it—that there is a profound, moral difference between the use of force for liberation and the use of force for conquest. You were here to liberate, not to conquer, and so you and those others did not doubt your cause. And you were right not to doubt.

You all knew that some things are worth dying for. One's country is worth dying for, and democracy is worth dying for, because it's the most deeply honorable form of government ever devised by man. All of you loved liberty. All of you were willing to fight tyranny, and you knew the people of your countries were behind you.

The Americans who fought here that morning knew

word of the invasion was spreading through the darkness back home. They fought—or felt in their hearts, though they couldn't know in fact, that in Georgia they were filling the churches at 4 A.M., in Kansas they were kneeling on their porches and praying, and in Philadelphia they were ringing the Liberty Bell.

Something else helped the men of D-Day: their rock-hard belief that Providence would have a great hand in the events that would unfold here; that God was an ally in this great cause. And so, the night before the invasion, when Colonel Wolverton asked his parachute troops to kneel with him in prayer he told them: Do not bow your heads, but look up so you can see God and ask His blessing in what we're about to do. Also that night, General Matthew Ridgway on his cot, listening in the darkness for the promise God made to Joshua: "I will not fail thee nor forsake thee."

These are the things that impelled them; these are the things that shaped the unity of the Allies.

When the war was over, there were lives to be rebuilt and governments to be returned to the people. There were nations to be reborn. Above all, there was a new peace to be assured. These were huge and daunting tasks. But the Allies summoned strength from the faith, belief, loyalty, and love of those who fell here. They rebuilt a new Europe together.

There was first a great reconciliation among those who had been enemies, all of whom had suffered so greatly. The United States did its part, creating the Marshall Plan to help rebuild our allies and our former enemies. The Marshall Plan led to the Atlantic Alliance—a great alliance that serves to this day as our shield for freedom, for prosperity, and for peace.

In spite of our great efforts and successes, not all that followed the end of the war was happy or planned. Some liberated countries were lost. The great sadness of this loss echoes down to our own time in the streets of Warsaw, Prague, and East Berlin. Soviet troops that came to the center of this continent did not leave when peace came. They're still there, uninvited, unwanted, unyielding, almost forty years after the war. Because of this, Allied forces still stand on this continent. Today, as forty years ago, our armies are here for only one purpose—to protect and defend democracy. The only territories we hold are memorials like this one and graveyards where our heroes rest.

We in America have learned bitter lessons from two world wars: It is better to be here ready to protect the peace, than to take blind shelter across the sea, rushing to respond only after freedom is lost. We've learned that isolationism never was and never will be an acceptable response to tyrannical governments with an expansionist intent.

But we try always to be prepared for peace; prepared to deter aggression; prepared to negotiate the reduction of arms; and, yes, prepared to reach out again in the spirit of reconciliation. In truth, there is no reconciliation we would welcome more than a reconciliation with the Soviet Union, so, together, we can lessen the risks of war, now and forever.

It's fitting to remember here the great losses also suffered by the Russian people during World War II: Twenty million perished, a terrible price that testifies to all the world the necessity of ending war. I tell you from my heart that we in the United States do not want war. We want to wipe from the face of the Earth the terrible weapons that man now has in his hands. And I tell you, we are ready to seize that beach-head. We look for some sign from the Soviet Union that they are willing to move forward, that they share our desire and love for peace, and that they will give up the ways of conquest. There must be a changing there that will allow us to turn our hope into action.

We will pray forever that some day that changing will come. But for now, particularly today, it is good and fitting to renew our commitment to each other, to our freedom, and to the alliance that protects it.

We are bound today by what bound us forty years ago, the same loyalties, traditions, and beliefs. We're bound by reality. The strength of America's allies is vital to the United States, and the American security guarantee is essential to the continued freedom of Europe's democracies. We were with you then; we are with you now. Your hopes are our hopes, and your destiny is our destiny.

Here, in this place where the West held together, let us make a vow to our dead. Let us show them by our actions that we understand what they died for. Let our actions say to them the words for which Matthew Ridgway listened: "I will not fail thee nor forsake thee."

Strengthened by their courage, heartened by their value [valor], and borne by their memory, let us continue to stand for the ideals for which they lived and died.

Thank you very much, and God bless you all.

The President spoke at 1:20 P.M. at the site of the Pointe du Hoc Ranger Monument where veterans of the Normandy invasion had assembled for the ceremony.

Following his remarks, the President unveiled memorial plaques to the 2nd and 5th Ranger Battalions. Then, escorted by Phil Rivers, superintendent of the Normandy American Cemetery, the President and Mrs. Reagan proceeded to the interior of the observation bunker. On leaving the bunker, the Reagans greeted each of the veterans.

Other Allied countries were represented at the ceremony by their heads of state and government: Queen Elizabeth II of the United Kingdom, Queen Beatrix of the Netherlands, King Olav V of Norway, King Baudouin I of Belgium, Grand Duke Jean of Luxembourg, and Prime Minister Pierre Trudeau of Canada.

Ronald Reagan's Speech at Omaha Beach

Commemorating the Fortieth Anniversary of the Normandy Invasion, June 6, 1984

Mr. President, distinguished guests, we stand today at a place of battle, one that forty years ago saw and felt the worst of war. Men bled and died here for a few feet of—or inches of—sand, as bullets and shellfire cut through their ranks. About them, General Omar Bradley later said, "Every man who set foot on Omaha Beach that day was a hero."

No speech can adequately portray their suffering, their sacrifice, their heroism. President Lincoln once reminded us that through their deeds, the dead of battle have spoken more eloquently for themselves than any of the living ever could. But we can only honor them by rededicating ourselves to the cause for which they gave a last full measure of devotion.

Today we do rededicate ourselves to that cause. And at this place of honor, we're humbled by the realization of how much so many gave to the cause of freedom and to their fellow man.

Some who survived the battle of June 6, 1944, are here today. Others who hoped to return never did.

"Someday, Lis, I'll go back," said Private First Class Peter Robert Zanatta, of the 37th Engineer Combat Battalion, and first assault wave to hit Omaha Beach. "I'll go back, and I'll see it all again. I'll see the beach, the barricades, and the graves."

Those words of Private Zanatta come to us from his daughter, Lisa Zanatta Henn, in a heart rending story about the event her father spoke of so often. "In his words, the Normandy invasion would change his life forever," she said. She tells some of his stories of World War II but says of her father, the story to end all stories was D-Day.

"He made me feel the fear of being on that boat waiting to land. I can smell the ocean and feel the seasickness. I can see the looks on his fellow soldiers' faces, the fear, the anguish, the uncertainty of what lay ahead. And when they landed, I can feel the strength and courage of the men who took those first steps through the tide to what must have surely looked like instant death."

Private Zanatta's daughter wrote to me: "I don't know

how or why I can feel this emptiness, this fear, or this determination, but I do. Maybe it's the bond I had with my father. All I know is that it brings tears to my eyes to think about my father as a twenty year old boy having to face that beach."

The anniversary of D-Day was always special for her family. And like all the families of those who went to war, she describes how she came to realize her own father's survival was a miracle: "So many men died. I know that my father watched many of his friends be killed. I know that he must have died inside a little each time. But his explanation to me was, 'You did what you had to do, and you kept on going.' "

When men like Private Zanatta and all our Allied forces stormed the beaches of Normandy forty years ago they came not as conquerors, but as liberators. When these troops swept across the French countryside and into the forests of Belgium and Luxembourg they came not to take, but to return what had been wrongly seized. When our forces marched into Germany they came not to prey on a brave and defeated people, but to nurture the seeds of democracy among those who yearned to be free again.

We salute them today. But, Mr. President, we also salute those who, like yourself, were already engaging the enemy inside your beloved country—the French Resistance. Your valiant struggle for France did so much to cripple the enemy and spur the advance of the armies of liberation. The French Forces of the Interior will forever personify courage and national spirit. They will be a timeless inspiration to all who are free and to all who would be free.

Today, in their memory, and for all who fought here, we celebrate the triumph of democracy. We reaffirm the unity of democratic peoples who fought a war and then joined with the vanquished in a firm resolve to keep the peace.

From a terrible war we learned that unity made us invincible; now, in peace, that same unity makes us secure. We sought to bring all freedom-loving nations together in a community dedicated to the defense and preservation of our sacred values. Our alliance, forged in the crucible of war, tempered and shaped by the realities of the postwar world, has succeeded. In Europe, the threat has been contained, the peace has been kept.

Today the living here assembled—officials, veterans, citizens—are a tribute to what was achieved here forty years ago. This land is secure. We are free. These things are worth fighting and dying for.

Lisa Zanatta Henn began her story by quoting her father, who promised that he would return to Normandy. She ended with a promise to her father, who died eight years ago of cancer: "I'm going there, Dad, and I'll see the beaches and the barricades and the monuments. I'll see the graves, and I'll put flowers there just like you wanted to do. I'll feel all the things you made me feel through your stories and your eyes. I'll never forget what you went through, Dad, nor will I let anyone else forget. And, Dad, I'll always be proud."

Through the words of his loving daughter, who is here with us today, a D-Day veteran has shown us the meaning of this day far better than any president can. It is enough for us to say about Private Zanatta and all the men of honor and courage who fought beside him four decades ago: We will always remember. We will always be proud. We will always be prepared, so we may always be free.

Thank you.

The President spoke at 4:33 P.M. at the Omaha Beach Memorial at Omaha Beach, France. In his opening remarks, he referred to President François Mitterrand of France.

Following the ceremony, President Reagan traveled to Utah Beach.

NOTES AND SOURCES

Numerous Reagan administration officials were interviewed for this book. Most helpful were James Baker III, Richard Darman, Michael Deaver, Anthony Dolan, Ken Duberstein, Bentley Elliott, Evan Galbraith, Peter Robinson, and Fred Ryan. In addition, letters from Lisa Zanatta Henn (December 30, 2004) and Anthony Dolan (January 18, 2005) were essential to my understanding the dynamics behind the Omaha Beach speech.

Being granted special access to Lisa Zanatta Henn's papers and home archive in Pleasanton, California, was also a major boon. Still a flight attendant for United Airlines, she has saved all of her father's World War II letters and her engaging correspondence with President Reagan. Her collection includes photographs and newspaper clippings.

Roger Kahn graciously allowed me access to his Reagan

file, which included three letters from the President on White House stationery.

Other important primary sources were:

Robert W. Black Collection, 1939–1991, U.S. Army Military History Institute, Carlisle, Pennsylvania. Contains material that Black gathered for his seminal books on the Rangers, including correspondence, unit histories, biographies, awards, and commendations.

Sims Gauthier Collection, Eisenhower Center for American Studies, University of New Orleans. This installation includes official documents from the planning of D-Day; of special interest are directives between the generals and their timetables for the attack.

Peter S. Kalikow World War II Oral History Archive, Eisenhower Center for American Studies, University of New Orleans. The largest collection of firsthand accounts from D-Day and Battle of the Bulge veterans. These remembrances from the men who were there made it possible for me to write this book. Especially helpful were those from Owen Brown, Eugene Elder, James Eikner, Ralph Goranson, Gerald Heaney, Albert Kamento, Jack Keating, Jack Kuhn, Louis Lisko, Leonard Lomell, Salva Maimone, and Frank South. All of their quotes used in this book come from these oral histories.

Army Heritage and Education Center, U.S. Army Military History Institute, Carlisle, Pennsylvania. A mother lode of material: unit histories, veteran questionnaires, and Army publications along with newspaper clippings, Army journal articles, and a vast collection of hard-to-find books.

National D-Day Museum, New Orleans, Louisiana. Opened on the fifty-sixth anniversary of D-Day and founded by my late friend and colleague Stephen E. Ambrose, this Smithsonian Institution affiliate honors not only the heroes of June 6, 1944, but all of the American stories of World War II.

James Earl Rudder Papers, Cushing Memorial Library, Texas A&M University, College Station, Texas. This massive collection of materials from Colonel Rudder's life proved invaluable for giving insight into his military career.

Ronald Reagan Presidential Library and Museum, Simi Valley, California. Four components of the Reagan Library proved an invaluable treasure trove. *The Presidential Papers* include the print versions of all of Reagan's speeches and public remarks along with all official paperwork regarding the hiring, promotion, and movement of White House staff, including Anthony Dolan, William Henkel, and Peggy Noonan. *The White House Daily Diary (Presidential Movements)* documents Reagan's hourly whereabouts during his presidency. It was particularly useful in tracking the chronology of his 1984 European trip. *The White House Office of Speechwriting (Research Records)* contains the primary documents relating to Reagan's D-Day speeches at Pointe du Hoc and Omaha Beach (Box 17033). Finally, *The White House Office of Speechwriting (Speech Drafts)* includes all of Anthony Dolan's and Peggy Noonan's speech drafts with their handwritten notes and marginalia, along with comments and suggestions from senior staff members and, eventually, Ronald Reagan himself.

Notes

The seizure of Pointe du Hoc by the U.S. Army Rangers' 2nd Battalion has been portrayed in such bestselling books as Cornelius Ryan's *The Longest Day: June 6, 1944* (New York: Fawcett, 1959) and Stephen E. Ambrose's *D-Day, June 6, 1944: The Climactic Battle of World War II* (New York: Simon & Schuster, 1994) and *The Victors: Eisenhower and His Boys, The Men of World War II* (New York: Simon & Schuster, 1998). All three were extremely helpful. A fine young military historian, JoAnna McDonald, wrote a small-press book titled *The Liberation of Pointe du Hoc* (Redondo Beach, Calif.: Rank & File, 2000), which I found valuable for its rigorous detail. It is by far the best available book on the Battle of Pointe du Hoc. JoAnna kindly provided me hard-to-find photographs, read a draft of the manuscript, and offered a number of helpful suggestions that were all followed.

The dean of Ranger history is Robert W. Black. His books—particularly his incisive *Rangers in World War II* (New York: Ivy Books, 1992)—were constant reference guides. Ronald J. Drez of New Orleans wrote about the Rangers in his seminal book *Voices of D-Day* (Baton Rouge: Louisiana State University Press, 1994), which was published on the fiftieth anniversary of the Normandy invasion. Drez, in fact, is responsible for locating and interviewing more 2nd Ranger Battalion veterans than any other scholar. He carefully reviewed the manuscript, for which I am extremely grateful.

More than any other books regarding the Reagan era, Peggy Noonan's *What I Saw at the Revolution* (New York: Random House, 1990) and *When Character Was King* (New York: Penguin, 2002) enabled me to properly understand the making of the "boys of Pointe du Hoc." Both of these works were extremely valuable. Peter B. Levy's *Encyclopedia of the Reagan-Bush Years* (Westport, Conn.: Greenwood Press, 1996) was a priceless reference guide.

Introduction: Setting the Stage

The story about Reagan's "statesmanship" quip comes from Peter Robinson's *How Ronald Reagan Changed My Life* (New York: Regan Books, 2003), pp. 250–52, as well as interviews with Mike Deaver, Evan Galbraith, and Peter Robinson. In looking at the events surrounding the bicentennial of the Battle of Yorktown I made use of Reagan's public papers, which include his remarks at a luncheon in Yorktown on October 18, 1981, and at a dinner that night honoring President Mitterrand in Williamsburg, along with his official speech honoring the bicentennial the next day in Yorktown. Helpful news accounts were Sarah Bird Wright, "The Victory of Yorktown," *Christian Science Monitor,* August 25, 1981; Howell Raines, "Reagan and Mitterrand Observe 200th Anniversary of Yorktown," *New York Times,* October 20, 1981; and Ben A. Franklin, "Even British Cheery at Fete of Yorktown," *New York Times,* October 20, 1981.

For Reagan's June 1982 trip to Europe I relied on the

White House Daily Diary and various newspaper clippings housed at the Reagan Library. Reagan's public remarks on this trip were referenced from *Public Papers of the Presidents of the United States: Ronald Reagan, 1982,* vol. 1 (Washington, D.C.: GPO, 1983): they include Reagan's radio address to the nation of June 5, 1982, his remarks that day commemorating the thirty-eighth anniversary of D-Day, a statement by Deputy Press Secretary Larry M. Speakes on June 6 in Paris, Reagan's statement following the conclusion of the Versailles summit (also on June 6), and his remarks following his meeting with Pope John Paul II at the Vatican on June 7.

Throughout this book I relied on Ambrose's *D-Day, June 6, 1944* for statistical information. Ambrose's scholarship in two other books also proved indispensable: *Citizen Soldiers: The U.S. Army from the Normandy Beaches to the Bulge to the Surrender of Germany, June 7, 1944–May 7, 1945* (New York: Simon & Schuster, 1994) and *The Supreme Commander: The War Years of General Dwight D. Eisenhower* (New York: Doubleday, 1969). Given the gargantuan nature of D-Day as a seminal historic event, there is no definitive set of casualty statistics. Also, the size of the Allied armada ranges from 5,000 to 7,000 ships, depending on the source.

Lou Cannon wrote "Reagan Hails D-Day Valor, Visits Graves," *Washington Post,* June 6, 1984, while Michael Dobbs penned "Normandy Braces Itself for Another Invasion," *Washington Post,* June 5, 1984. I also made use of Peter Almond's June 5, 1984, *Washington Times* story, "D-Day Memories." A truly moving early account of a D-Day veteran bringing a family member back to Normandy is W. C. Heinz, "I Took My Son to Omaha Beach," *Collier's,* June 11, 1954.

Jon Meacham called FDR a "national pastor" in *Franklin and Winston: An Intimate Portrait of an Epic Friendship* (New York: Random House, 2003), p. 284. The Ambrose quote comes from *Citizen Soldiers,* p. 472, and Michael Barone's is from *Our Country: The Shaping of America from Roosevelt to Reagan* (New York: Simon & Schuster, 1990), p. 641.

The Tom Brokaw commentary is from an October 16, 2004, interview.

In recent years a number of good books on D-Day have been published. While not directly quoted in this work, they've all been quite useful: Martin Gilbert, *D-Day* (New York: John Wiley & Sons, 2004), Robin Neillands, *The Battle of Normandy, 1944* (New York: Cassell, 2002), John C. McManus, *The Americans at Normandy: The Summer of 1944—The American War from the Normandy Beaches to Falaise* (New York: Forge Books, 2004), and Flint Whitlock, *The Fighting First: The Untold Story of the Big Red One on D-Day* (New York: Westview, 2004). Of constant help was Barrett Tillman's *Brassey's D-Day Encyclopedia: The Normandy Invasion, A–Z* (Washington, D.C.: Brassey's, 2004).

1: Darby's Rangers

The history of the early Rangers has been fully documented in a number of important books. My knowledge of their antics derives mainly from John R. Cuneo, *Robert Rogers of the Rangers* (New York: Oxford University Press, 1959); Walter Webb, *The Texas Rangers,* 2nd ed. (Boston: Houghton Mifflin, 1965); and John K. Mahon, "Anglo-American Methods of Indian Warfare," *Mississippi Valley Historical Review* 45

(1958), pp. 254–75. For the Rangers during the Civil War, see *Mosby's War Reminiscence* (New York: Pageant, 1958). Articles used for this chapter include "Hard-Hitting Commandos," *New York Times Magazine,* April 5, 1942; and Bruce Thomas, "The Commando," *Harper's,* March 1942 (JoAnna McDonald brought these articles to my attention). The story of Truscott's mission to Northern Ireland can be found in L. K. Truscott Jr., *Command Missions: A Personal Story* (New York: E.P. Dutton, 1954); his quote about the name "Rangers" is on p. 40. A good introduction to Darby, his men, and their actions during World War II is Mir Bahmanyar, *Darby's Rangers, 1942–45* (Wellingborough, Eng.: Osprey, 2003).

Of great help in better understanding Ranger history was Jerome J. Haggerty, "A History of the Ranger Battalions in World War II," Ph.D. diss., Fordham University, 1982. For biographical data of various Rangers mentioned, see Henry Henderson, *Colonel Jack Hays: Texas Ranger* (San Antonio: Naylor, 1954); and T. R. Fehrenbach, *Commandoes: The Destruction of a People* (New York: Knopf, 1974); Colonel G. F. R. Henderson, *Stonewall Jackson and the American Civil War* (New York: Longmans, Green, 1955); James W. Williamson, *Mosby's Rangers* (New York: Sturges and Walton, 1909); and Donald B. Chidsey, *The War in the South* (New York: Crown, 1969).

Much has been written on William Darby, but the most riveting work is James Altieri's *The Spearheaders* (Indianapolis: Bobbs-Merrill, 1960). I consulted his memoir, coauthored with William H. Baumer, *Darby's Rangers: We Led the Way* (San Rafael, Calif.: Presidio, 1980), when writing about World War II Rangers. Important factual information was gleaned from Gordon L. Rottman, *U.S. Army Rangers and LRRP Units, 1942–87* (Oxford: Osprey, 1987); and especially Black, *Rangers in World War II*.

There are a number of fine books about the Dieppe debacle. I relied on Torence Robertson, *Dieppe: The Shame and the Glory* (Boston: Atlantic/Little Brown, 1962); Norman L. R. Franks, *The Greatest Air Battle: Dieppe, 19th August, 1942* (London: Kimber, 1979); Will Fowler, *Commandos at Dieppe: Rehearsal for D-Day* (London: HarperCollins, 2002); and Black, *Rangers in World War II*, pp. 26–45.

For Churchill on Dieppe—including references to "hunter class," "butcher and bolt," and striking companies"—see Black, p. 5. The Rangers' exploits in Africa are outlined in ibid., pp. 47–76.

2: Rudder's Rangers

For statistics pertaining to the all-important training facility of Camp Forrest, I relied on an official history provided by the Arnold Engineering Development Center at Arnold Air Force Base, which now occupies the area. Also helpful was George M. Clark, William Weber, and Ronald Paradis, *2nd Ranger Battalion: The Narrative History of Headquarters Company, April 1943–May 1945* (Carlisle, Pa.: Army Military History Institute, n.d.). Again, Rottman's *U.S. Army Rangers and LRRP Units* and McDonald's *The Liberation of Pointe du Hoc* guided the writing of this chapter.

For inspiration on James E. Rudder, I relied on Ronald Lane's riveting *Rudder's Rangers* (Guilderland, N.Y.: Ranger Associates, 1979) and Sam Blair's informative "Earl Rudder:

The Greatest Aggie Hero," *The Eagle*, August 16, 1996 (found in the James Earl Rudder Papers). The uplifting story about Private Petty is from Lane's book.

For the history of Fort Pierce I contacted the Fort Pierce Inlet State Park and referenced Robert A. Taylor's illustrated history, *World War II in Fort Pierce* (Charleston, S.C.: Arcadia, 1999). The stories about Herman Stein come from McDonald, *The Liberation of Pointe du Hoc.*

I learned of the 2nd Ranger Battalion's setting the two-hour speed march record from James Ladd, *Commandos and Rangers of World War II* (New York: St. Martin's, 1978), p. 265.

Omar Bradley recounted his conversation with Rudder in *A Soldier's Story* (New York: Henry Holt, 1951), p. 269. For information on Andrew Higgins, I drew from my "The Man Who Won the War for Us," *American Heritage*, May–June 2000; and Jerry Strahan, *Andrew Jackson Higgins and the Boats That Won World War II* (Baton Rouge: Louisiana State University Press, 1998).

Lieutenant G. K. Hodenfield's story, "I Climbed the Cliffs with the Rangers," appeared in the *Saturday Evening Post* on August 19, 1944; it included the photograph that graces the cover of this book.

3: Climbing the Cliffs, Destroying the Guns

This chapter is based largely on the accounts in the Peter S. Kalikow World War II Oral History Archive at the Eisenhower Center. My more recent telephone conversations with Leonard Lomell and Jack Kuhn (who died in 2004) also helped me flesh out various details. The book I wrote with Ronald J. Drez, *Voices of Valor: D-Day, June 6, 1944,* was useful in crafting this chapter. Drez, a decorated combat veteran of the Vietnam War, also reviewed this section for errors (he also had conducted most of the Eisenhower Center oral history interviews cited).

Also integral to the writing of this chapter was the all-important "after-battle report," which was filed on July 22, 1944. A copy was provided by the U.S. Army Military History Institute. The institute was the source as well for the extremely helpful report about the taking of the cliffs, "Small Unit Actions: Rangers: Pointe du Hoc: 2nd Ranger Battalion, June 6, 1944."

Jonathan Bastable's *Voices from D–Day* (Newton Abbot, Devon: David & Charles Publishers, 2004) includes a section (pp. 138–45) that deals exclusively with the Battle of Pointe du Hoc. Margaret Rudder is quoted in Sam Blair, "Earl Rudder: The Greatest Aggie Hero," *The Eagle*, August 1996.

A wealth of information on the *Texas* was furnished by the Battleship *Texas* State Historic Site in LaPorte, Texas.

Frank South's quotations come from Patrick K. O'Donnell, ed., *Beyond Valor: World War II's Ranger and Airborne Veterans Reveal the Heart of Combat* (New York: Free Press, 2001).

I found the essential military documents from James Rudder's distinguished military career, such as his Silver Star citation, in his papers at the Cushing Memorial Library, Texas A&M University.

4: Reagan's Hollywood War

The major events in Reagan's pre-presidential years are delineated in "Ronald Reagan's Pre-Presidential Time Line, 1911–1980," made available by the Reagan Library. Much has been written about Reagan's youth and early manhood in the Midwest. I relied principally on Lou Cannon, *Governor Reagan: His Rise to Power* (New York: Public Affairs, 2003); Garry Wills, *Reagan's America: Innocents at Home* (Garden City, N.Y.: Doubleday, 1987); Ronald Reagan, *An American Life* (New York: Simon & Schuster, 1990); and Anne Edwards, *Early Reagan: The Rise to Power* (New York: William Morrow, 1987), especially pp. 3–37.

For the World War II years, I drew upon Reagan's own account in *An American Life*, pp. 88–120. My information about Reagan's military service comes from his 120-page "Military Record" file, housed at the Reagan Library; it includes a detailed chronology of his obligations and accomplishments from 1935 to 1945. Also helpful in this regard was Cannon, *Governor Reagan,* pp. 66–73.

The story about the B-17 pilot came from Wills, *Reagan's America*. Chapter 19 of Wills's book, "War Movies," offers a brilliant analysis of Reagan's penchant for sugarcoating or fabricating World War II stories. Wills covers such Reagan tall tales as witnessing the Jewish concentration camps himself and portraying the U.S. Armed Forces as racially integrated at the time of Pearl Harbor.

Fort Mason is now a cultural and educational institution called the Fort Mason Center. The staff there kindly provided me with the history of this legendary Bay Area military base.

The Edmund Morris quote about "celluloid commandos" comes from his *Dutch* (New York: Random House, 1999), p. 196.

A great deal has been written about Reagan's film career. My main source was Tony Thomas, *The Films of Ronald Reagan* (Secaucus, N.J.: Citadel Press, 1980), pp. 16–142; the Irving Berlin anecdote comes from this book. Also helpful were Stephen Vaughan, *Ronald Reagan in Hollywood* (Cambridge: Cambridge University Press, 1994); Bill Boyarsky, *The Rise of Ronald Reagan* (New York: Random House, 1968); Charles Higham, *Warner Brothers* (New York: Charles Scribner's Sons, 1975); and Frank McAdams, "What Price Glory, Captain Reagan?," *Los Angeles Times,* August 27, 1980.

Lou Cannon quotes Hugh Heclo's "sacramental vision of America" in *Governor Reagan,* p. 118, citing Heclo's paper "Ronald Reagan and the American Public Philosophy," presented at a May 27–30, 2002, conference at the University of California–San Diego. For the "diction" quote, see Wills, pp. 192–200.

Historian David McCullough's quotes appear in William E. Leuchtenberg, *In the Shadow of FDR: From Harry Truman to Bill Clinton,* rev. ed. (Ithaca, N.Y.: Cornell University Press, 1993), pp. 209–35. For Reagan on glimpsing FDR in a parade, see Kiron K. Skinner, Annelise Anderson, and Martin Anderson, *Reagan: A Life in Letters* (New York: Free Press, 2003).

Mary Beth Brown's analysis of Roosevelt's effect on Reagan appears on p. 59 of her *Hand of Providence: The Strong and Quiet Faith of Ronald Reagan* (Nashville: WND Books, 2004). Other helpful books that dealt with Reagan's relationship with God were Michael K. Deaver, *A Different Drummer: My Thirty Years with Ronald Reagan* (New York: HarperCollins, 2001); Dinesh D'Souza, *How an Ordinary Man Became an Extraordinary Leader* (New York: Free Press, 1997); Michael Reagan, *On the Outside Looking In* (New York: Kensington, 1988); and Robinson, *How Ronald Reagan Changed My Life*. Reagan himself wrote an interesting religious article, "My Faith," which appeared in *Modern Screen* in May 1950.

Matthew Dallek's *The Right Moment* (New York: Free Press, 2000) traces Reagan's conversion to the Republican Party and is a valuable source for his rise in California politics—particularly the importance of his speech at the 1964 GOP convention in San Francisco.

The Hollywood Beverly Christian Church story and Reagan's interest in FDR's "arsenal of democracy" come from Cannon, *Governor Reagan*, pp. 62–81.

For the history of speechwriting, I referred to Bradley H. Patterson, *The White House Staff: Inside the West Wing and Beyond* (Washington, D.C.: Brookings Institution Press, 2000), pp. 162–72; the Dewey quote comes from this source.

5: Peggy Noonan Prepares for Pointe du Hoc

Noonan's masterly *What I Saw at the Revolution* was indispensable in writing this chapter; most of her quotes come from this memoir. Other published sources used are Peggy Noonan, "The Education of Dan Rather," *Wall Street Journal*, December 2, 2004; "Q&A: Peggy Noonan," *Atlanta Journal-Constitution*, June 20, 2004; and Ann Morse, "Meeting Peggy Noonan," *Crisis* 22, no. 8 (September 2004), pp. 34–39.

For Reagan's love of quoting scripture, see Michael Deaver, *Behind the Scenes* (New York: William Morrow, 1987), and his *A Different Drummer.* Three books that Noonan clearly read for background information were Cornelius Ryan's *The Longest Day,* Jean Compagnon's *The Normandy Landing: The Strategic Victory of the War* (Rennes: Éditions Ouest-France, 1979), and John Keegan's *Six Armies in Normandy: From D-Day to the Liberation of Paris* (New York: Viking, 1982).

My three or four conversations with Ben Elliott helped me understand the protocol of the Reagan speechwriting team. I also interviewed Michael Deaver on November 16, 2004, and William Henkel on November 23, 2004. A great reference on the inner workings of the executive branch is Patterson, *The White House Staff.*

Good biographical articles on Darman are "Richard Darman Returns to Kennedy School Faculty," *The Harvard University Gazette,* January 15, 1998; John K. Mashek, "Richard Darman: The Man Behind the Numbers," *Boston Globe,* January 30, 1990; and Martin Walker, "The Budget Makers and the Breaker," *The Guardian* (London), October 9, 1990. For the Normandy speech being like a "screenplay," see Frank Rich, "First Reagan, Now His Stunt Double," *New York Times,* June 13, 2004.

I was able to contact Colonel Ray Stakes at his home in Ocean Springs, Mississippi. He kindly sent me his U.S. Army record and a July 2, 1984, letter that President Reagan wrote to Lieutenant General E. Grange Jr.

As noted in the text, Lance Morrow's "Every Man Was a Hero," *Time,* May 28, 1984, helped trigger the fortieth-anniversary hoopla. For a full explanation of how the Robert Capa D-Day photograph came about, see Robert Capa, *Photographs* (New York: Aperture/Philadelphia Museum of Art, 1996), pp. 100–101.

During the two months leading up to the fortieth anniversary of D-Day, Reagan's speechwriting staff collected all of the significant D-Day articles that appeared in newspapers and periodicals around the country; they're now archived in the White House Office of Speechwriting, Research Records, housed at the Reagan Library. Among the most useful to me were Richard L. Stout, " 'What Awe!' One Reporter Recalls the D-Day Invasion," *Christian Science Monitor*, April 18, 1984; Thomas H. Wolf, "D Day Remembered: 'A Brief Afterglow of Battle Survived,' " *Smithsonian*, May 1984; Richard C. Firstman, "A Jump into Past Glory," *Newsday*, May 8, 1984; Mark S. Smith, "D-Day's Secret Tragedy, Five Weeks Before the Invasion," Associated Press, May 13, 1984; Jack Dorsey, "D-Day Volunteers Plan Fortieth Reunion," *Norfolk Virginian-Pilot*, May 12, 1984; Stanley Meisler, "D-Day—A Legacy of Gratitude," *Los Angeles Times*, May 13, 1984; Marty Bishop, "D-Day: Allied Forces Played Waiting Game," *Pentagram* (Washington, D.C.), May 17, 1984; Stephen Webbe, "D-Day Revisited," *Christian Science Monitor*, May 21, 1984; John Vinocur, "French Are Said to Give a D-Day Rebuff to West Germany's Leader," *New York Times*, May 21, 1984; Paul Beeman, "D-Day's 'First on the Beach'—I Was Just Doing My Job," *New York Post*, May 24, 1984; Storer Rowley, "D-Day of 1984: Americans Ready to Invade Again," *Chicago Tribune*, May 27, 1984; Charles E. Claffey, "We Have Not Forgotten,"

Boston Globe, May 27, 1984; and Jack Schnedler, "France Ready for Another Invasion," *New York Post,* May 29, 1984. Besides Lance Morrow's piece, *Time*'s May 28 issue included the articles " 'Every Man Was a Hero,' " "Daisies from the Killing Ground," and "Overpaid, Oversexed, Over Here."

Other pre-D-Day fortieth-anniversary articles found in Noonan's file include Michael Cordts, "Again, He'll Cover D-Day," *Chicago Sun-Times,* June 1, 1984; Ben Wattenberg, "D-Day Remains a Dominant Symbol for the Western World," *Newark Star-Ledger,* June 1, 1984; Glen W. Martin, "WWII's Fateful Decisions," *San Antonio Express,* June 2, 1984; Andy Rooney, "Return to Normandy Brings Back Memories," *Colorado Springs Gazette-Telegraph,* June 2, 1984; John F. Burns, "As Allies Recall D-Day, Russians Say It Was Just a Sideshow to Their War," *New York Times,* June 2, 1984. The following all date from June 3, 1984: Scott Thurston, "D-Day: America's Young Men Fought for Freedom, and Found Glory" and "USS Augusta Cruised to the Fore," *Atlanta Journal-Constitution;* Jingle Davis, "Paratroopers Jumped to the Wind with Moxie of 'Screaming Eagles,' " *Atlanta Journal-Constitution;* Joseph W. Grigg, "Thunder of War Fed Londoners' Hope for Freedom," *Atlanta Journal-Constitution;* W. R. Higginbotham, "Forty Years Later, a Memory of D-Day Remains as Fresh as Yesterday," *Philadelphia Inquirer;* Fred Rasmussen, "Back Tracks," *Baltimore Sun;* Robert Ruby, "Forty Years Later," *Baltimore Sun Magazine;* G. K. Hodenfield, "Victory at 'Atlantic Wall,' " *Norfolk Virginian-Pilot;* George DeWan, "The Shadow of D-Day," *Newsday;* Allen Pusey, "D-Day: The World Remembers," *Dallas Morning News;* Sheila Taylor,

"What D-Day Means to Europe's Pepsi Generation," "Invasion Brought Freedom," and "The Staging Area," *Dallas Morning News;* Hal Parrett, "Normandy Invasion: The Allies Prepare," *Dallas Morning News;* Maxwell D. Taylor, "The Battles on Omaha and Utah Beaches," *Hartford Courant;* and William Cockerman, "D-Day Memories Still Vivid After Forty Years," *Hartford Courant.* The *Chicago Sun-Times* carried dozens of stories in its "D-Day Special" that day, including John H. Thompson, "Hitting the Beach!"; Michael Cordts, "The Bloody Crusade That Changed History"; Tony Ginnetti, "Operation Overlord: A Chronology"; Lloyd Green, "The Big Plane That Couldn't"; Max Hastings, "D-Day: 'An Encounter Between Flesh and Fire' "; Marcia Schnedler, "Illinois Had Its Share of Heroes on Longest Day"; and Michael Cordts, "The Women They Married" and " 'Dear Mrs. De Witt: Your Son's Loss Was a Shock.' " Also useful were Sheila Taylor, "Woman Recalls American Visitors to First Liberated Town," *Dallas Morning News,* June 4, 1984; and Brad Bradley, "The Day Before D-Day," *Dallas Morning News,* June 5, 1984. The file also has transcriptions from the CBS and NBC evening news broadcast segments that previewed the D-Day events on June 2, 1984.

6: Reagan's Normandy Day

Once again, most of the Noonan-related material comes from *What I Saw at the Revolution.* The numerous drafts of Noonan's speech have been meticulously organized at the Reagan Library. My interviews with Ben Elliott in November and December 2004 helped me understand the role Noonan played on the White House speechwriting team.

The opening quote in this chapter is from Lord Mountbatten's foreword to James Ladd, *Commandos and Rangers of World War II* (New York: St. Martin's Press, 1978). The number of honors Rudder's Rangers netted is from the U.S. Army Military History Institute, Carlisle, Pennsylvania.

A number of documents found at the Reagan Library inform this chapter, including: "Press Briefings, John O. Marsh, Secretary of the Army on the Normandy Portion of the President's Trip to Europe, May 22, 1984" (transcript); Kim White (speechwriting office), White House memorandum, "Speeches to Normandy Advance Office," June 1, 1984; Robert M. Kimmitt (National Security Council), memorandum, "President's Draft Speech for Omaha Beach Ceremonies," May 28, 1984; William F. Martin (NSC), memorandum, "Annotated Agenda for Europe," April 1984; Robert M. Kimmitt (NSC), memorandum, "Pointe du Hoc Presidential Remarks," May 24, 1984; William F. Martin (NSC), memorandum to Ben Elliott and Kim Timmons, "Annotated Agenda for the Trip to Europe," April 19, 1984; Robert C. McFarlane, memorandum for the President, "Your Trip to Europe," April 16, 1984 (includes capsule biographies of various Rudder's Rangers); and Julie Case, undated memorandum to Peggy Noonan [May 1984]. Case suggested that Noonan highlight the story of Thomas D. Howie, 116th Infantry Regiment, 29th Infantry Division. He had been killed in action near Saint-Lô, France, on July 17, 1944, after having won the Silver Star, Bronze Star, Purple Heart, Croix de Guerre with Palm, and the Légion d'Honneur.

Especially helpful was "Preserving Peace and Prosperity: The President's Trip to Europe," June 1984, found in White

House Office of Speechwriting, Research Records, Box 17033. Len Lomell's wife provided the capsule biographies of surviving 2nd Battalion Rangers that are in this box. In the same box at the Reagan Library, I found Noonan's memo to Elliott of May 30, 1984.

My December 23, 2004, interview with Roger Kahn, as well as the letters and clippings he provided, informed my understanding of the role *The Boys of Summer* (New York: Harper & Row, 1972) played in the writing of the Pointe du Hoc speech. David Leeming's *Stephen Spender: A Life in Modernism* (New York: Henry Holt, 1999) helped me better understand the work of this great British poet. Noonan had read up on Spender in *Current Biography Yearbook* (New York: H.W. Wilson, 1977) and in books on modern poetry by Louis Untermeyer.

The Lisa Zanatta Henn and Peter Zanatta correspondence and documents discussed in this chapter are also found in White House Office of Speechwriting, Research Records, Box 17033. Included in this file are Lisa Zanatta Henn to President Ronald Reagan, March 15, 1984; Peter Zanatta's military record (and honorable discharge); Colonel M. P. Caulfield to Lisa Zanatta Henn, May 10, 1984; and Reagan's undated letter suggesting that Zanatta Henn's travel expenses be picked up by the U.S. government. Private Zanatta wrote a number of letters home, all of which are in Lisa's home archive; the ones I found of particular interest for this book were written on August 3 and August 9, 1943; April 9, October 6, and December 9, 1944; and July 9, 1945.

For more information about Lorraine Wagner and her

long epistolary relationship with Reagan, see my own "Ronald Reagan's Pen Pal," *The New Yorker,* July 26, 1999.

Margaret Thatcher's memoir *The Downing Street Years* (New York: HarperCollins, 1993) was enlightening on the Reagan administration's diplomatic relations with Great Britain.

Herman Stein's reclimbing the cliffs came from Michael Dobbs, "Reagan to Visit Site of Ranger Assault; Aged Veteran Climbs Cliff to Relive D-Day Drama," *Washington Post,* June 6, 1984.

For the Wintz family, see Joan Burney, "Fifty Years Later, Family Sees Cliffs D-Day Survivor Once Scaled," *Omaha World-Herald,* September 6, 1994.

For Reagan's visit to Omaha Beach, see Lou Cannon, "Reagan Hails D-Day Valor, Visits Graves," *Washington Post,* June 7, 1984; and Steven R. Weisman, "Soviet Union's Role in War Acknowledged by Reagan," *New York Times,* June 7, 1984.

For color commentary about Reagan's day at Normandy, see John Vinocur, "D-Day Plus Forty Years," *New York Times,* May 13, 1984; and Donnie Radcliffe, "Of War and Memories," *Washington Post,* June 7, 1984.

Conversations and interviews with Bill Plante, Dan Rather, Walter Cronkite, and Sam Donaldson helped me better understand the U.S. media perspective on Reagan's speeches.

Much has been written about Theodore Roosevelt Jr.'s heroism at D-Day. In his diary, General George Patton Jr. noted that TR Jr. was the "bravest soldier" he ever knew. See H. Paul Jeffers, *Theodore Roosevelt, Jr.: The Life of a War Hero* (Novato, Calif.: Presidio, 2000).

For information about the Omaha Beach area memorials, I consulted various pamphlets published by the American Battle Monuments Commission in 1975.

The Pointe du Hoc speech was reprinted on June 7, 1984, in the *New York Times*, p. A12, and the *Washington Times*, p. 5A.

The White House Daily Diary, Presidential Movements, housed at the Reagan Library, provides Reagan's exact movements on June 6, 1984. (There is a one-hour time difference between Great Britain [LDT] and France [FDT].)

LDT 1142	**Depart Winfield House [London] via Marine One**
FDT 1402	**Arrive Pointe du Hoc Landing Zone, Normandy, France**
1503	**Depart Pointe du Hoc Landing Zone via Marine One**
1520	**Arrive Omaha Beach Landing Zone**
1700	**Depart Omaha Beach Landing Zone via Marine One**
1711	**Arrive Utah Beach Landing Zone**
1900	**Depart Utah Beach Landing Zone via Marine One**
LDT 1916	**Arrive Winfield House**

7: After the Speeches

The correspondence between Ronald Reagan and Lisa Zanatta Henn comes from Lisa Zanatta Henn's collection in Pleasanton, California. Frank Rich cited Reagan's "blockbuster elegy" in "First Reagan, Now His Stunt Double," *New York Times*, June 13, 2004. Elbert E. Scudero's letter of June 18, 1984, was forwarded to me by Lisa Zanatta Henn.

For more on the Bitburg debacle, see Ilya Levkov, *Bitburg and Beyond: Encounters in American, German and Jewish History* (New York: Shapolsky, 1987); Tom Shales,

"On the Air," *Washington Post,* May 6, 1985; William Drozdiak, "Bitburg: Furor Embitters City That Loved Its Americans," *Washington Post,* May 6, 1985; and Harrison E. Salisbury, "Again, Signs of Presidential 'Second-Term Blues,'" *New York Times,* May 12, 1985. Reagan explained his Bitburg experience in *An American Life*, pp. 376–84. Also, I quoted from Ronald Reagan to Lisa Zanatta Henn, May 22, 1985.

A transcript of Reagan's January 11, 1989, Farewell Address was furnished by the Reagan Library.

Roger Kahn provided an interview on December 23, 2004, as well as Reagan's letter to him of July 26, 1988. For more on *The Winning Team,* see Thomas, *The Films of Ronald Reagan,* p. 193.

A fine book on U.S.-Soviet diplomacy in the 1980s is Jack F. Matlock Jr., *Reagan and Gorbachev: How the Cold War Ended* (New York: Random House, 2004). The story of Reagan's famous Berlin speech comes from Robinson's *How Ronald Reagan Changed My Life,* pp. 85–113, and interviews with Peter Robinson and Tony Dolan. Some of the material on Reagan's Berlin Wall address was adapted from my essay "A Clarion Call for Freedom," in *Hoover Digest,* no. 4 (1999).

Epilogue

For the USS *Eisenhower* anecdote, see Steven R. Weisman, "Soviet Union's Role in War Acknowledged by Reagan," *New York Times,* June 7, 1984. For information pertaining to the Eisenhower-Reagan relationship, see Stephen E. Ambrose, *Eisenhower: Soldier and President* (New York: Simon

& Schuster, 1990, p. 572). The quote about Eisenhower's need for approval comes from Peter Lyon, *Eisenhower: Portrait of the Hero* (Boston: Little, Brown, 1974), p. 304. In *An American Life,* Reagan relates how in 1952 he sent Eisenhower a telegram urging him to run for president as a Democrat (p. 133).

A good analysis of Reagan and the neoconservatives is Wills, *Reagan's America,* pp. 404–5.

The death of Ronald Reagan created hundreds of obituaries and retrospectives on his life. Some of the best, which informed my thinking, were: Marilyn Berger, "Ronald Reagan Dies at Ninety-three," *New York Times,* June 6, 2004; Martin Kasindorf, "A Nation Remembers," *USA Today,* June 7, 2004; David Von Drehle, "Reagan Hailed as Leader for 'the Ages,'" *Washington Post,* June 12, 2004; Jon Meacham, "Portrait of a President," *Newsweek,* June 14, 2004; and Nancy Gibbs, "The All-American President," *Time,* June 14, 2004.

Special thanks to the CNN team in Normandy with whom I worked on June 5–6, 2004; also to the producers I served with as historical commentator for NBC/MSNBC News coverage of Reagan's funeral.

ACKNOWLEDGMENTS

Once again my editor at William Morrow, Claire Wachtel, is responsible for my getting this book out in a timely fashion. With great fortitude, and insight, she guided this project along from start to finish. She is a dear friend and patient counselor. Others at William Morrow who were helpful include Claire's assistant, Kevin Callahan, as well as Michael Morrison, Lisa Gallagher, Ben Bruton, Kim Lewis, Aryana Hendrawan, and Shubhani Sarkar. I look forward to working with them on other books.

William Morrow arranged for the services of Trent Duffy, a masterly line editor. Trent helped make my prose clearer and smoother at the same time that he carefully combed every paragraph for factual errors and fixed my typos. This is the third book we've worked together on and I trust him wholeheartedly.

My staff at the Eisenhower Center for American Studies at the University of New Orleans was once again extremely helpful. Lisa Weisdorffer rode shotgun over the entire project while Andrew Travers proved to be an indefatigable research assistant. Both of them have now moved with me to Tulane University's Theodore Roosevelt Center for American Civilization; we are a team. Michael Edwards—a longtime Eisenhower Center scholar—helped me locate both World War II veterans and a few important documents.

Working closely with the Ronald Reagan Library in Simi Valley, California, was a wonderful experience. The institution is filled from top to bottom with first-rate individuals. Special mention, however, goes to the smart and fastidious Lisa Jones. Always genial, she delivered the proper documents for my research needs in record time. Steven Branch and Ray Wilson were very helpful in tracking down photos for the book.

At the U.S. Army Military History Institute at Carlisle, Pennsylvania, Richard Baker was a constant source of knowledge. As was Tom Herring, my wise friend from the U.S. Army Ranger Association. Military historian JoAnna McDonald of San Diego, whose book *The Liberation of Pointe du Hoc* helped me in so many ways, was a constant source of historical information and personal encouragement. At NBC News, where I serve as an historical commentator, Tom Brokaw, Elena Nachmanoff, Tammy Haddad, and Peter Giordano helped me locate individuals, footage, and photographs in a prompt fashion. Further fact-checking assistance was provided by Len Lomell, Lisa Zanatta Henn, Roger Kahn, Michael Edwards, Frank South,

Steve Maguire, Bentley Elliott, Marty Morgan, James Baker III, Peter Robinson, Anthony Dolan, Andy Ambrose, George Stevens Jr., and Michael Deaver.

On a more personal note, this book is dedicated to Ron Drez, a colleague of mine first at the University of New Orleans and now at Tulane University. He is a true American hero.

And to my beautiful wife, Anne, and our two young children, Benton and Johnathan. The three of them make my life worth living.

INDEX

BOOKS BY DOUGLAS BRINKLEY

THE GREAT DELUGE *(New in Hardcover, Spring 2006)*

Hurricane Katrina, New Orleans and the Mississippi Gulf Coast

ISBN 0-06-112423-0 (hardcover) • 0-06-112894-5 (CD)

The complete tale of the terrible storm, offering a unique, piercing analysis of the ongoing crisis, its historical roots, and its repercussions for America.

THE WORLD WAR II MEMORIAL

A Grateful Nation Remembers

ISBN 0-06-085158-9 (paperback)

A distinguished, beautifully illustrated companion volume to the newly opened World War II Memorial on the National Mall.

PARISH PRIEST

Father Michael McGivney and American Catholicism

ISBN 0-06-077684-6 (hardcover) • 0-06-085340-9 (CD) • 0-06-085348-4 (large print)

An in-depth biography of Father Michael J. McGivney, the Roman Catholic priest who stood up to anti-Papal prejudice in America and founded the Knights of Columbus.

TOUR OF DUTY *(New York Times* Bestseller*)*

John Kerry and the Vietnam War

ISBN 0-06-056529-2 (paperback) • 0-06-058976-0 (large print)

Covering more than four decades, this is the first full-scale, definitive account of Kerry's journey from war to peace.

THE WORLD WAR II DESK REFERENCE

With the Eisenhower Center for American Studies

ISBN 0-06-052651-3 (hardcover)

For World War II enthusiasts, history buffs, and anyone interested in our nation's history, this is the one book to own.

HarperCollins*Publishers*